Mechanotransduction

Mechanotransduction

Cell Signaling to Cell Response

PAUL A. SUNDARAM
University of Puerto Rico, Mayagüez, Puerto Rico

ELSEVIER
ACADEMIC PRESS
An imprint of Elsevier

Academic Press is an imprint of Elsevier
125 London Wall, London EC2Y 5AS, United Kingdom
525 B Street, Suite 1650, San Diego, CA 92101, United States
50 Hampshire Street, 5th Floor, Cambridge, MA 02139, United States
The Boulevard, Langford Lane, Kidlington, Oxford OX5 1GB, United Kingdom

British Library Cataloguing-in-Publication Data
A catalogue record for this book is available from the British Library

Library of Congress Cataloging-in-Publication Data
A catalog record for this book is available from the Library of Congress

ISBN: 978-0-12-817882-9

For Information on all Academic Press publications
visit our website at https://www.elsevier.com/books-and-journals

Publisher: Mara Conner
Acquisitions Editor: Fiona Geraghty
Editorial Project Manager: Fernanda A. Oliveira
Production Project Manager: Prasanna Kalyanaraman
Cover Designer: Mark Rogers

Typeset by MPS Limited, Chennai, India

Dedication

My wife, Fabiola, who is my perfect complement,
my true companion, without whom I would be incomplete.

My children, Paola Alejandra and Julian Patrick, who are my crown
and joy and hopefully will carry the torch forward.

My parents, Chandran and Vimala, who taught me the important
things in life.

Contents

Preface

Mechanotransduction is a complex topic which involves a lot of biology but also mechanical forces and deformations. A concise textbook dedicated to this topic is unavailable which could bring together complex topics such as cell signaling and cell response for the engineering student and assemble it with the concepts of mechanical stress and strain which would be difficult to grasp for students in biology. Of course, the field of mechanotransduction is rife with research articles which are challenging for early graduate students in engineering who have little or no background in biology. It dawned on me that a textbook on mechanotransduction if written in a simple manner would be a valuable tool for students in both engineering and biology. This book will be a starting point for students to grasp the basic concepts of mechanotransduction and focus on specific review articles to get a more in-depth knowledge.

The book is written is such a fashion so as to introduce the concept of mechanotransduction in Chapter 1, Introduction as well as some of its practical implications in the human body. This is followed by the fundamental principles of signal transduction in cells including signal reception, transduction, and response in Chapter 2, Cell signaling: an overview. Chapter 3, Cell signaling under mechanical stimulus in specific cell types delves into the effect of mechanical stimulus on various cell types. Attention is paid to some specific cell types which have been studied in some detail in literature. Chapter 4, Cell characteristics resulting from mechanical stimulus deals with changes in cell characteristics under the influence of mechanical stimulus including the effect on stem cells, cancer cells, cell proliferation, and remodeling of the extracellular matrix. Mechanotransduction in tissues is described in Chapter 5, Tissue mechanotransduction with the goal of understanding the influence of mechanical stimulus in larger groups of cells or tissues. The practical applications of mechanotransduction in actual clinical treatments through mechanotherapy are presented in the Chapter 6. Chapter 7, Future thrusts in the field of mechanotransduction suggests some topics which are worth researching in further understanding the broad theme of mechanotransduction.

Each chapter is written in a form that it can be treated independently but there is a connection between the chapters also. At the end of most of the chapters there are a few questions under "review and project" to enable the student to research and think further about some of these topics. It is my hope that this book brings the concepts of mechanotransduction in a clear and simple manner to the student to build a firm foundation upon which more complex concepts can be constructed.

Paul A. Sundaram

Acknowledgments

The desire to write a textbook that would be accessible to students motivated me after I first taught a graduate course in cell mechanics in 2012. To teach the course, I had to use material from a variety of sources including books, research articles, and reviews. I decided to write a book on mechanotransduction since a textbook on this topic is currently unavailable. Writing this book was a major project and many persons were directly or indirectly involved from conceptualizing it to its birthing. When Elsevier approached me with the idea of writing a book, I took the bait, never for a moment imagining the many challenges that this project would bring along. I was convinced to come up with a title and a table of contents. When I first produced the table of contents, the book began to take shape. The Elsevier Publishing team quickly tied me into a contract from which there was no turning back. During the course of writing the sample chapter, the publishing editor Fiona Geraghty passed the baton on to Fernanda Oliveira who has been instrumental in getting this book out. The Elsevier team worked with me, hand in hand, to ensure timely delivery. The book in finally a reality for which I am grateful to Elsevier.

Of course, the impetus to write on this topic definitely came from my graduate students at the University of Puerto Rico, Mayaguez. Each one of them, through their painstaking, forward-thinking research have somehow made me think more profoundly about the phenomenon of mechanotransduction and how to succinctly relay complex concepts to students with clarity and simplicity. It would not be superfluous to emphasize that my students were the drivers for this output. I am, of course, grateful to the Department of Mechanical Engineering at the University of Puerto Rico at Mayaguez for always providing me the space and support throughout my career.

Finally, the writing of this book became a family project. I am grateful to my wife and children for their patience when I was absorbed and sometimes obsessed with writing to produce a textbook palatable for the student. My wife and daughter created, designed, and produced many of the illustrations in the manuscript, without which this book would not have been finished. I cannot but be grateful to the King of kings and Lord of lords to whom all glory is due.

Foreword

Have you ever stopped to ponder how the body works? More specifically, how, as a complex machine, does the body develop and sustain integrated form and function of all its working tissue and organ components? To address such questions, one must embrace mechanobiology, a field that operates at the interface of biology, engineering, and physics to describe how physical forces at the molecular, cellular, and tissue levels guide tissue development, physiology, and disease. Fundamental to mechanobiology is mechanotransduction, the process by which cells sense and respond to physical stimuli by converting them to downstream biochemical signals. Paul A. Sundaram's book, *Mechanotransduction: Cell Signaling to Cell Response*, brings together the latest developments in the field of mechanobiology, providing a rich foundation on the fundamental cell machinery and associated mechanotransduction processes involved in health and disease. This includes definition of the underlying mechanisms by which various cell types modulate their characteristics and behaviors in response to (1) physical forces (e.g., fluid shear, tension), (2) stiffness of the surrounding extracellular scaffold, and (3) connectivity with other cells, with referenced applications specific to tissue and organ systems, including musculoskeletal, cardiovascular, pulmonary, and nervous. The latter chapters, focus on how fundamental mechanobiology and mechanotransduction principles are inspiring design and development of engineered therapeutics, devices, and computational simulations, with applications in medicine. This book is designed for readers from all relevant fields, including engineering, life sciences, physical sciences, and computational sciences. It provides an excellent reference and current update for those new to or currently working within the mechanobiology arena. This book represents a timely production and is highly recommended to all.

Sherry L. Harbin
Weldon School of Biomedical Engineering and
Department of Basic Medical Sciences, Purdue University,
West Lafayette, IN, United States

CHAPTER 1

Introduction

The trillion or so cells in the human body are divided into functional groups, which act synchronously and in a highly coordinated manner in performing the body's physiological functions. All these cells are controlled by means of signals, which are stimuli of various types, which provide the specific impetus necessary to elicit the corresponding function. Many of these stimulators are chemical in nature although the wide impact and the role of physical parameters, including mechanical forces and factors in realizing routine functions, cannot be denied. In biological terms, simply put, in all if not most processes, there is a reception of the signal, an appropriate transduction, which is many times unknown, and a resulting action, which brings about a response from the target whether it is an organ or a tissue or a cell. For example, clearly, there is some chemical signaling occurring through neurotransmission, which causes both voluntary and involuntary muscle movement. The obvious physical movement of our limbs and other parts of the body is accomplished by a complex set of signals and transduction. However, when we look at more subtle mechanical stimuli as initiators of normal physiological function, there is no denying the added involvement of the mechanical forces exerted by the cardiac pump in circulating the large volume of blood throughout the body. The flow of blood exerts a mechanical force on the blood vessels whose cells are constantly subjected to a stretching and relaxing force variations. Peristalsis is the undulating movement of the muscles and involves the transport of food from the mouth to the stomach through mechanical forces. The absorption of nutrients at the intestines along this trajectory and finally excretion accomplished through bowel movement is also very much a mechanical aspect of our bodily functions. Normal breathing is yet another example where a cyclic force is imposed on cells that line the lung tissue with precision and consistency. Mechanical forces become especially important in embryonic development during the gastrulation phase. The sensitivity of most, if not all, cells in the human body to physical or mechanical forces is undeniable. We will look at more specific examples about individual cells and cell groups or tissues and understand the particular roleplay of mechanical stimuli and the corresponding transduction to accomplish physiological functions in greater detail in the chapters that follow.

While gross physiological functions such as kinesis or cardiac function are obvious facts in relating and appreciating the importance of mechanical forces in the human body, there are many more subtle effects that escape our attention. Protein expression and conformation are very much controlled by mechanical factors at the cellular level. In fact, the absence or deficiency of physical parameters in cellular functions has

Mechanotransduction
DOI: https://doi.org/10.1016/B978-0-12-817882-9.00002-1

resulted in a number of pathological conditions, disease, and dysfunction. The role of mechanical forces in organogenesis is increasingly evident, and scientists have slowly realized the utmost significance of the mechanical interaction between a cell and its microenvironment, including neighboring cells as well as the extracellular matrix.

While the interest in the effect of mechanical factors on cells, tissues, and organs is undeniable, only in recent years has there been an uptrend in the interest in understanding these phenomena. The effects of mechanical stimuli have been reported in the literature for quite a while, but their critical roles in the proliferation of cells, cell differentiation, motility, protein synthesis, etc. are only being recently researched. The mechanisms underlying the positive and adverse effects on cells are being looked at from the fundamental perspective; ideas on these are proposed based on experimental evidence and reported. From the early publication of D'Arcy Thompson, scientists have reported various efforts to understand the effects of the mechanical manipulation of cells. However, a firm reference regarding the influence of mechanical forces at the cellular level gained strength only in the 50s. Since then, our understanding of the interplay between biological events influenced by mechanical forces has steadily increased with critical discoveries resulting from the development of innovative experimental techniques. This concept of mechanochemical transduction (mechanical stimuli producing biochemical responses) and its role in a number of physiological and pathological conditions gave further impetus to the study of this phenomenon.

From the early discovery that cancer cells could become adherent and spread on soft agar to the mechanosensing capability of the plasma membrane, it has become evident that physical forces play an important role in cell physiology. The tension differential between the inner and outer cell membrane layers in erythrocytes, which leads to changes in cell shape, was an important step in understanding cell migration. Armed with the experimental data and techniques to report evidence of ion channel activation by shear stress and the discovery of the integrin family of proteins, the 1980s opened the field of mechanotransduction to a new level. With the development of sophisticated experimental techniques, the understanding of the mechanisms such as integrin-mediated mechanotransduction became more accessible. The innovative technique of Traction Force Microscopy enabled the measurement of small-scale forces that cells apply on their substrates, and the use of Atomic Force Microscopy (AFM) to research the unfolding of proteins followed by the ability over the years to delve even deeper into miniscule structures such as the response of talin to mechanical stimuli has given us a more profound understanding of the effect of mechanical stimuli on biological responses in cells. AFM was skillfully utilized to stretch proteins and uncover domains that heretofore unknown. More recent techniques that have enabled the researcher to obtain nanoscale resolution to track the stretch—relaxation cycles of proteins. Novel fluorescence techniques exposed the presence of molecules such as paxillin, talin, focal adhesion kinase, etc. in the focal adhesion complexes. With such advances, it is completely feasible to quantify cell-scale forces to a high level of precision and study this parameter at different scales. In the latter half of the first decade of the 21st century, the first fluorescence resonance energy transfer-based protein stretch sensor was developed. In vivo techniques gained popularity, as live-cell imaging of talin, vinculin, and other connectors helped our understanding of mechanosensors. Somewhere in this historical time frame, the term *mechanotransduction* was introduced to describe the cell physiology process whereby cells convert mechanical stimuli into biological responses. Innovative experimentation and newly developed techniques have greatly helped us to understand the essence of mechanical stimuli on cells, tissues, and organs in that hierarchical order. Thus, it has become possible to understand the effect of mechanical forces and the corresponding cell response in an even more profound manner. Fig. 1.1 is a historical timeline of the development of the study of *mechanotransduction* and the significant milestones reached in this process. The recent development of sophisticated instrumentation has enabled discovery and observation, identification, and description of several intracellular molecular mechanisms. Substrates with submicron post-like structures were fabricated to allow for the measurement of cell stress. More recently, cell division has been hypothesized to be controlled by molecular motors, which operate in conjunction with the microtubule-organizing centers to bring about this function. Also, a report about constant communication between the endoplasmic reticulum and mitochondria through local transport was recently highlighted.

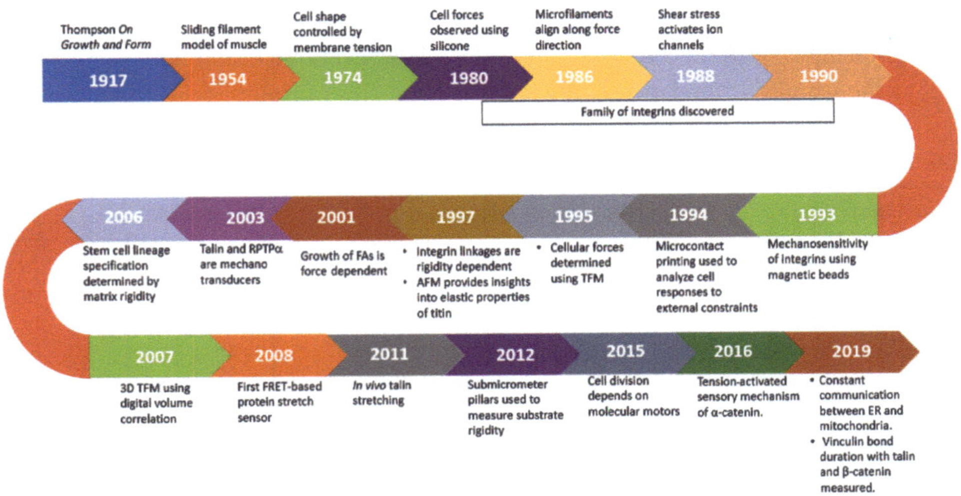

Figure 1.1 A timeline of important milestones including discoveries and development of instrumentation, which has led to the phenomenal growth of the field of mechanotransduction in the last few decades. *AFM*, Atomic Force Microscopy; *FRET*, Förster Resonance Energy Transfer or Fluorescence Resonance Energy Transfer; *TFM* Traction Force Microscopy; *RPTP*; Receptor Protein Tyrosine Phosphatase. *Data taken from Iskratsch T, Wolfenson H, Sheetz MP. Appreciating force and shape-the rise of mechanotransduction in cell biology. Nat Rev Mol Cell Biol 2014;15(12):825–33.*

The term mechanotransduction has now been generally accepted to define the conversion of mechanical signals into biochemical or biomechanical responses in the cell and further extended to include tissues and organs. As in other cell signaling mechanisms, mechanotransduction involves a sensory or reception phase, a transduction phase, and a resulting action in and by the cell. Stretch-sensitive ion channels and the interaction involving the extracellular matrix and the cytoskeleton have now been accepted as standard mechanosensors or receptors and a number of intracellular signaling pathways have been identified, which regulate gene expression and protein synthesis. These, in turn, are critical to cell metabolism or physiological function. The importance of these functions in system operation involving organ function cannot be overemphasized.

There is no denying the fact that the wide gamut of events further exacerbates the difficulty in pinpointing just one or two mechanotransduction processes as controlling all these occurrences. Rather, a combination of such mechanotransducers must be accounted for to accomplish such a complex array of physiological events. A certain specific mechanical stimulus can provoke the operation of various signaling mechanisms and accomplish a variety of corresponding physiological functions. On the other hand, various types of mechanical stimuli can result in the same cell response.

Precise actions at the microscale are normally required to cause large-scale changes at the tissue, organ, or even the system level. Table 1.1 shows estimated values of mechanical stimuli on various cells and highlights the rather small values necessary to drive large-scale changes in the corresponding tissue or organ. In the majority of physiological changes, protein expression is ultimately responsible, and many signaling cascades at the cellular level depend on conformal changes in the protein structure.

Table 1.1 Estimated magnitudes of mechanical stimuli on cell in vivo.

Physical profile	Mechanical stimulus	Magnitudes
Arterial blood flow	Fluid shear stress	1–3 Pa
Cell migration	Traction stress	3–5.5 kPa (normal cells)
		1 kPa (cancer cells)
Proximal tubule flow	Fluid shear stress	0.3 Pa
	Fluid drag force	0.0074 pN/microvillus
	Bending torque	0.016 pN-mm/microvillus
Stretch-activated channels	Membrane tension	0.012 N/m
Outer hair cell stereocilia	Compression stiffness	0.01 N/m
	Force/D membrane potential	0.02 0.1–20 pN/mV
Bone canaliculi	Fluid shear stress	0.8–3 Pa
	Fluid drag force	20 × shear force
	Tissue strain	0.03%–0.1%

Source: Adapted from Orr AW, Helmke BP, Blackman BR, Schwartz MA. Mechanisms of mechanotransduction. Develop Cell 2006;10(1):11–20.

From a thermodynamic perspective, these conformal alterations must be accompanied by a decrease in free energy of the reaction or process. In extending this argument to processes involving mechanical stimuli, the changes in protein conformation must support the consequent decrease in free energy, failing which a process will not occur. For example, an increase in tension within the phospholipid bilayer will result in the activation or opening of stretch-activated ion channels located in cell membrane provided that the change in surface area will result in a decrease in free energy.

Thus the important role of mechanotransduction in these developmental processes cannot be understated, and the drastic changes in cell movement, which occur during embryogenesis during gastrulation and neurulation, must be accompanied by a free energy decrease. It appears that a feedback loop from the forces exerted on the tissues during embryogenesis guides tissue movement. It is speculated that mechanotransduction resulting from fluid flow controls the formation of left—right asymmetry during embryonic development. Also, the morphology of the cardiovascular system is highly influenced by the mechanical stimuli exerted by the blood flow via pressure on smooth muscle cells in the lumen and shear stress on the endothelial lining. Bone tissue is shaped by the forces applied on it by muscles.

In the following chapters, a sequential but exclusive development of the general topic of mechanotransduction is provided with the additional detail necessary to delve deeper into the subject. As a caveat, mechanotransduction, in one sense, is still a fledgling theme with important implications in human health. Chapter 2, Cell Signaling: An Overview, provides a fairly detailed description of cell signaling including the trifecta of reception, transduction, and protein synthesis. Chapter 3, Cell Signaling Under Mechanical Stimulus in Specific Cell Types, deals with cell signaling in specific cell types in response to mechanical stimulus, followed by changes in the characteristics of normal and cancerous cells and the behavior of stem cells in Chapter 4, Cell Characteristics Resulting From Mechanical Stimulus. Mechanotransduction at the tissue level and its effect on organ function will be dealt with in Chapter 5, Tissue Mechanotransduction. The role of mechanical forces in disease evolution as well as concepts in mechanotherapy to exploit the positive effects of mechanical loads and strains on human health will be highlighted in Chapter 6, Mechanotherapy. The last chapter will shed some light on future opportunities to further study this emerging and exciting field of mechanotransduction.

Review and Project

1. Write a list of 4-5 situations where mechanical stimulus is applied in the human body to bring about physiological function.
2. For 2 or 3 of the above cases, indicate the stimulus, the possible signal transduction in general terms and the response.

CHAPTER 2

Cell signaling: An overview

2.1 Introduction

The purpose of a signal in a cell is to accomplish some type of physiological function. Hence, the cell has to receive a signal, which then results in transduction within the cell and ultimately causes the cell to perform its function. In a nutshell, the overall mechanism in cell communication, including the initiation and response to a signal, consists of three distinct phases or steps: (1) signal reception, (2) signal transduction, and (3) signal response. The cell receives a signal through ligand binding with a receptor protein, which has both an extracellular portion to capture the signal and an intracellular or cytosol part, which transmits the signal to other intracellular proteins in a process called transduction. The signal is finally received by effector proteins, which accomplish the physiological function of the cell as the response. In what follows, a bird's-eye view of the cell signaling process is provided without attempting to get into details of advanced biochemical reactions that occur in these processes, which are beyond the scope of this book. However, sufficient information is provided for the reader to grasp the fundamentals of cell signaling. Advanced students are directed to study this phenomenon in many research articles that are available, which deal specifically with this topic.

2.2 Modes of cell communication

Cell signaling entails the transmission of information from one cell to another cell or a group of cells. Hence, cells must be able to communicate with each other to pass along this information onward. Faulty communication between cells results in deficiency in the transmission of information, which, in the human body, may lead to pathologies or disease. In a complex multicellular system such as the human body, it is imperative that there is effective communication between different cell groups for proper functioning of the system. It should be noted that there is a redundancy in the signaling, resulting in the same cell function, in case the primary signal is unable to accomplish its objective because of a deficiency in that signaling pathway. There is a plethora of communication pathways between cells and groups of cells. Cells respond to a variety of signals, which can be classified as chemical and mechanical stimuli. In a general sense, all cells are subjected to some kind of signaling, which elicits a physiological response. If we think about a simple sensation such as hunger, the initial signal comes from the pituitary gland, which in and of itself is triggered by a lack of energy production in the body. This signal is followed by the secretion of gastric juices in the

Mechanotransduction
DOI: https://doi.org/10.1016/B978-0-12-817882-9.00004-5

stomach, which causes a "burning" sensation, creating an urge in us to consume food. The feedback from the energy generated by the breakdown of sugars from the food that was ingested results in the signal being shut off until more energy is needed.

When an animal, let us say, for example, a lioness, hunts for her prey, the sensation of hunger drives her to identify the prey, stalk the victim, hunt, and kill it. Although this appears to be a simple process, one can imagine at the cellular level, the many types of cells that are activated by an initiator. In this case, the hormone epinephrine is the trigger, which provides increased energy for the lioness to chase her prey, overpower it, and then deprive it of life. During this process, a complex interaction and activation of different cell groups with specific functions provides a heightened sense of hearing, sight, ability to accelerate to high speeds, and brute strength to kill the prey just through the release of epinephrine from the adrenal glands. The same adrenaline rush is the effector for the prey to run faster and be aware of its hunter, with the goal of escaping the lioness and preserving its life. In human beings, the same sensation is experienced when we feel we are in danger, as epinephrine is released into our bloodstream. Compare these survival instincts in special circumstances with the normal activity of a person who has to artificially create this heightened state during strenuous physical exercise. In this case, the release of epinephrine is coerced by the intense level of physical activity imposed on the human body. It is not surprising then that to accomplish these highly complex tasks, communication between groups of cells becomes critical to the healthy physiological functioning of the human body.

Cell-to-cell communication is accomplished in a variety of ways depending on the proximity of the signaling cell to the target cell, as well as the function that the system aims to accomplish. The simplest communication path is through autocrine signaling where the cell initiating the signal and the responding cell are one and the same. Another method of communication between cells is through the direct contact of the initiating cell with another target cell. The signal initiating cell comes into contact with the target cell and secretes the chemical agent (signal) that is necessary to accomplish the required function. This signal transmission is enabled by means of intercellular connections called gap junctions through the exchange or passage of small molecules. Gap junctions are transmembranal structures that connect adjacent cells and enable the transport of small molecules and ions (particularly Ca^{2+}) between neighboring cells. The induction of B cells, which come in direct contact with helper T cells at the threat of antigens in our immune system, is an example of cell communication by direct contact. The secretion of neurotransmitters in the synaptic junctions is an example of another type of cell communication called paracrine signaling where the signaling cell and the target cell are in close proximity to each other but not in direct contact. Fig. 2.1 is a simple illustration of the release of neurotransmitters in the synaptic junction from the presynaptic axon and further transmission of this communication down the line to the proximal postsynaptic axon. Such type of signal is an example of cell communication, in this case, specifically, paracrine signaling, generated frequently in the gastrointestinal tract.

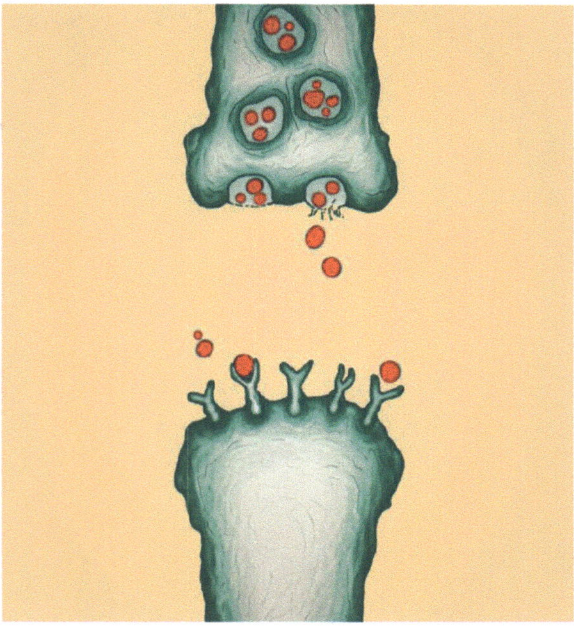

Figure 2.1 Typical cell communication. Simple illustration of typical cell–cell communication. In the example shown, the release of neurotransmitters from one axon into the synaptic junction of the neighboring axon results in the transmission of a cell signal.

The production of hormones by the pituitary gland to accomplish a physiological function far away from the point of release is a third cell communication method known as endocrine signaling. In this case, the hormone, which is the signaling molecule, is released into the bloodstream and travels a relatively long distance to trigger an appropriate response in another cell group much farther away. The distance effect on cell signaling is summarized in Fig. 2.2.

2.3 Phases in cell signaling

The cell signaling process can be divided into three distinct phases, which occur in the following order, reception, transduction, and response. This means that transduction cannot take place before reception and neither can response occur before signal reception and transduction.

2.3.1 Reception

Cell signaling is normally initiated when a small molecule or even a larger protein is released by one set of cells to trigger a response. This initiator signal, commonly called a ligand, attaches to a cell receptor, as seen in Fig. 2.3. This signaling molecule is

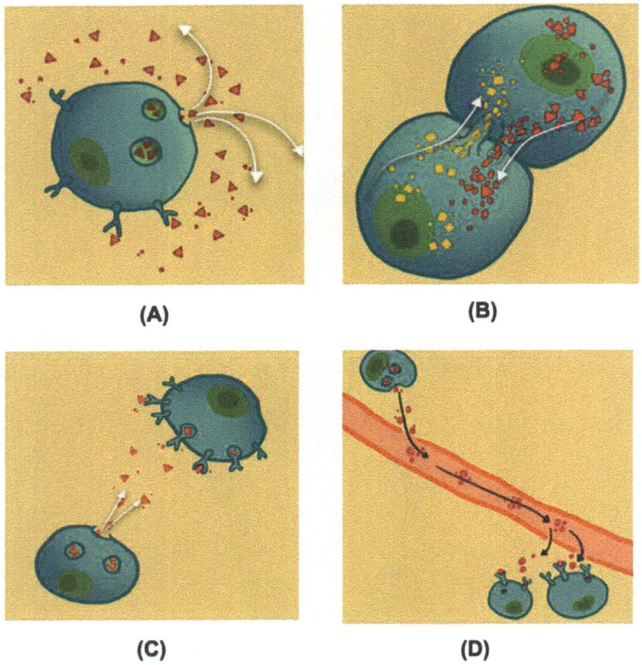

Figure 2.2 Cell communication methods. (A) Autocrine signaling illustrating the release of ligands by a cell to be received by its own receptor and accomplish the physiological function of the cell. (B) During direct contact signaling, the cells are in contact with each other where the effector cell secretes the required chemical signal into the target cell. (C) Paracrine signaling, indicating close proximity of effector and target cells where ligands from the effector cells are released for binding with receptors in the target cells. (D) Endocrine signaling, where the effector and target cells are at a considerable distance from each other, and the ligand is transported long distances to elicit a response from the target cell.

sometimes called a first messenger. The receptor is responsible for receiving the signal and is normally a transmembranal protein. This phase of cell communication is called reception. Fig. 2.3 further illustrates that only a specific ligand can attach to a specific receptor to activate it. The analogy is similar to a lock and key. Only a specific key (ligand) can open a particular lock (receptor) under normal operating conditions and the ligand attaches to a specific binding site on the receptor to activate it. Cell receptors can be grouped into transmembranal and intracellular receptors based on their location. The transmembranal receptors are positioned in the cell membrane as part of the phospholipid bilayer with an extracellular component to receive the signal from the ligand and a cytosol part, which facilitates the transmission of the signal inside the cell. Intracellular receptors are either found in the cytoplasm and later enter the nucleus or are situated inside the cell nucleus. The receptor will be activated and remain so for as long as it is necessary and then become inactive until is needed again.

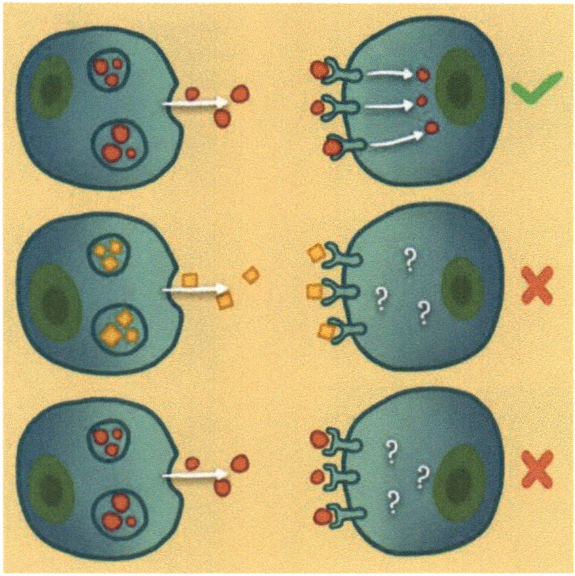

Figure 2.3 Activation of receptor by ligand binding.

A specific ligand can only attach to a corresponding specific receptor. This specificity is critically important in not activating incorrect signaling pathways, which can result in undesirable diseases and other pathologies. The illustration emphasizes the specificity of the ligand and receptor for effective cell signaling. Schematically, the green checkmark corresponds to the correct ligand that is needed to activate the signaling process.

The ability of a ligand to activate a receptor is based on the specific shapes of both ligand and receptor. This conformal property ensures that only a specific ligand can effectively attach to a specific receptor and cause the activation of the cell. This is also a protection mechanism for the cell so that cell processes, which could cause pathological problems or other diseases, are not inadvertently activated. Besides, the activation process is transient so that this mechanism is not operating continually and is not prolonged in keeping with the limited life cycle of the cell.

There are four types of cell receptors, of which three are transmembranal and one is intracellular. These can be seen in Fig. 2.4. The illustration is a graphical description of the mechanisms of each of these receptors. The details of these mechanisms are given in a later section. The three transmembranal receptors are as follows:

1. Ligand-gated channel receptors, also called ion-channel receptors, which are activated by ligands that are hydrophilic and hence not soluble in lipids. This type of receptor undergoes conformal change when activated and allows for sodium ingress and potassium egress in the cell. The ligands, in this case, are small molecules, and the activation process occurs in a matter of milliseconds.

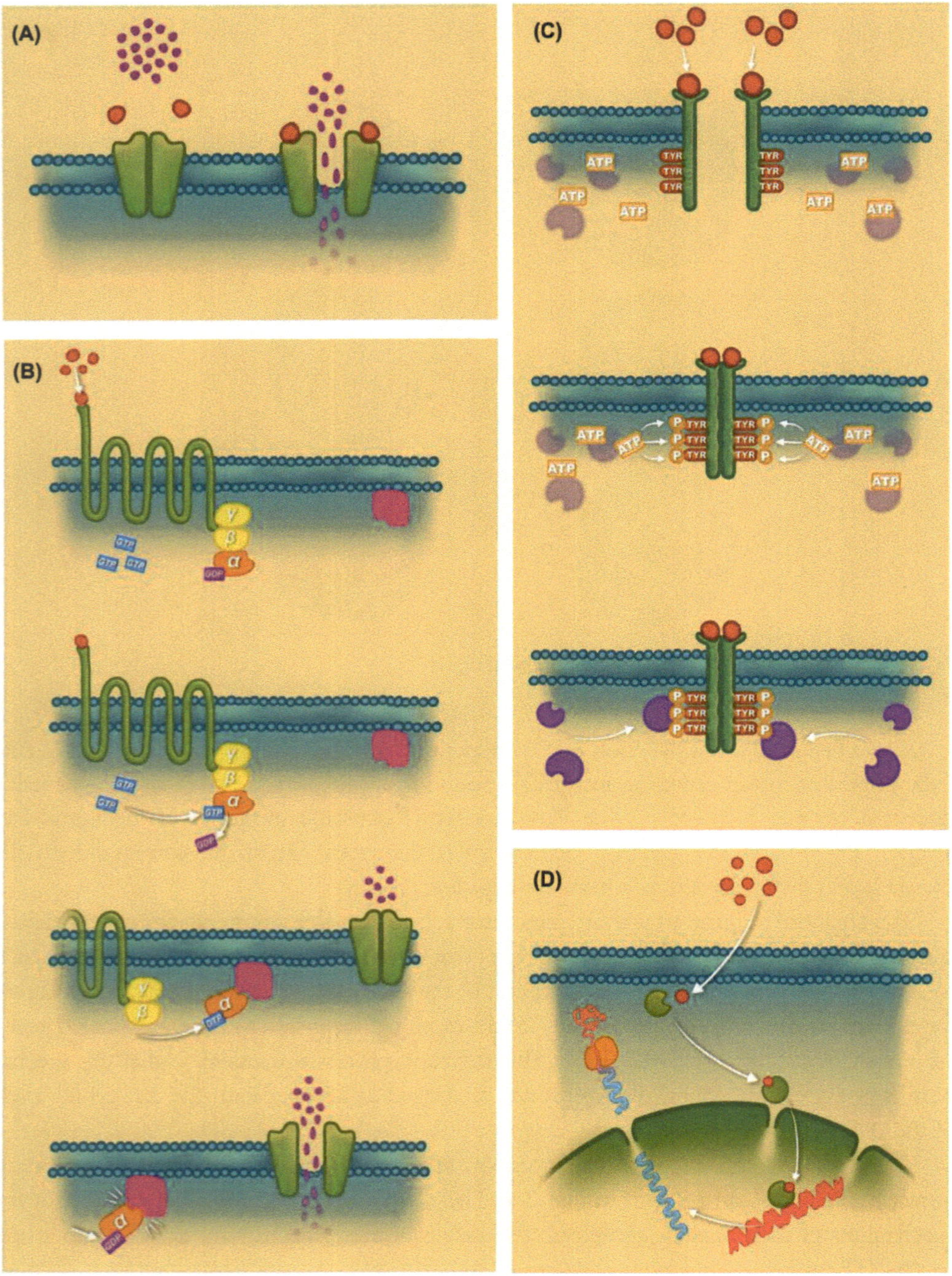

Figure 2.4 Types of cell receptors. *ATP*, Adenosine triphosphate; *GDP*, guanosine diphosphate; *GTP*, guanine triphosphate.Transmembranal receptors such as (A) ligand-gated channel receptor allow the passage of nutrients by selective opening under stimulus by a specific ligand molecule, (B) G protein–coupled protein receptor and corresponding mechanism of operation, (C) receptor tyrosine kinase activated by ligands, and (D) intracellular receptors, which are hydrophobic and able to penetrate the cell membrane to reach the nucleus. The corresponding mechanisms for each of these receptors are clearly illustrated.

2. G protein—coupled receptors (GPCRs), form the largest group of receptors, number about 800 and are activated by small-molecule ligands. These receptors also undergo conformal changes and, in turn, activate other enzymes or proteins inside the cell. These are also known as seven transmembranal receptors because these have seven alpha-helical structures traversing the phospholipid layer in a serpentine shape. The activation process takes a matter of seconds for these receptors.
3. Enzyme-linked receptors have kinases linked to their structure and are activated by large protein molecules. One group of these receptors is called receptor tyrosine kinases (RTKs). Activation of these receptors results in enzymes being activated within cells, a process that may take a few hours to occur. One typical activation is when guanine triphosphate (GTP) is transformed to cyclic adenylyl monophosphate (cAMP), which, in turn, triggers the function of other signaling cascades inside the cell.

   Intracellular receptors, as the name suggests, are located inside the cell. The ligands for these receptors are hydrophobic and are hence able to penetrate the phospholipid bilayer of the cell membrane.
4. Intracellular receptors are located in the cytoplasm and later translocated to the nucleus or may be found directly inside the nucleus. Activation of these receptors also takes hours, which affects transcription factors and hence protein expression. As these are located inside the cell, their ligands are lipid-soluble molecules, which traverse the cell membrane to initiate signaling. Steroids are typical examples of ligands of this type.

2.3.2 Transduction

The conversion of the external stimulus sensed through the cell receptor into an internal signal is called transduction. The details of this transduction process are quite complex. In many ways, it is relatively easy to study the stimulus and the response, although the signal transduction process itself is not wholly transparent. The transmission of the signal in the cell is described by signaling pathways, which indicate a sequence of biochemical reactions ultimately leading to an appropriate response from the cell. There is a myriad of signaling pathways available through which the cell can perform its functions, and the complex interaction of these mechanisms may be difficult to understand. However, some of the more common mechanisms will be mentioned to provide a flavor of signal transduction and all that it entails in normal and healthy physiological activity of the cell. Once the signal is received by ligand binding to a cell receptor, it undergoes a conformal change, and the signal is now transduced in the cytosol through a number of messengers called effectors. These intracellular messengers are usually other proteins that pass the signal further along to other proteins called secondary messengers or relay molecules and so forth until the information is received by the cell nucleus where gene transcription takes place. This process of signal transmission is normally referred to as a signaling pathway for a simple transmission or signaling cascade for more complex activity involving many messengers. Finally, the nucleus sends forth a response, which could be associated with the cell's metabolic activity, simply by expressing or producing a protein or by activating cytoskeletal elements in the cell to perform a mechanical function such as migration. Fig. 2.5 summarizes the process of signal transduction in a succinct manner.

2.3.2.1 Some common signaling pathways

Generally, the signal transmission inside the cell is performed via on/off type switches, which activate the required proteins in the signaling pathway. This is accomplished by the phosphorylation and dephosphorylation of the messenger proteins. The switches that normally turn on phosphorylation are enzymes called kinases, whereas those that dephosphorylate or turn off protein function are called phosphatases. The protein is activated by the catalytic action of the enzymes, resulting in the addition of a phosphate to the guanosine diphosphate (GDP) or adenosine diphosphate (ADP) to convert it to GTP or adenosine triphosphate (ATP), respectively, and it is turned off when the GTP or ATP is hydrolyzed back to the GDP or ADP state. Some of the more common signaling pathways are ion channel signaling, GPCR signaling, inositol triphosphate (IP_3) signaling, RTK signaling, and mitogen-activated protein kinase (MAPK) signaling, which will be explained in some more detail in the following sections. The reader is referred to Fig. 2.4 for a graphical representation of these signaling mechanisms/pathways.

2.3.2.1.1 Ligand-gated ion channel signaling pathway

This signaling pathway is also activated by the binding of a ligand to the receptor protein located on the extracellular side in the plasma membrane. This allows the flow of small ions such a Na^+, K^+, or Ca^{2+} from the extracellular space to the cytoplasm and vice versa, altering ion concentrations within the cell. Na^+, K^+, or Ca^{2+} ion concentrations are extremely important for various cell functions, especially in the transmission of signals in nerve cells. Ligand action in opening the ion channel to selectively allow the transport of Na^+ ions inside the cell is critical for cell homeostasis. Many cellular functions are heavily dependent on the Ca^{2+} concentration in the cytosol. Ca^{2+} is a very important second messenger and is involved in cell division and in muscle movement. There are three classes of ligand-gated ion channels: P2X, glutamate, and cys-loop channels, with a variety of signaling pathways. The ion channel gate remains closed when it is not activated by a specific ligand, and the ions can no longer enter or leave the cell. Cell signaling based on the ligand-gated ion channels is very important in the transmission of electrical signals in neurons. When vesicles laden with neurotransmitters from one nerve cell are released in the synaptic cleft of the neighboring neuron, these ligands bind to the receptor in the postsynaptic dendrite,

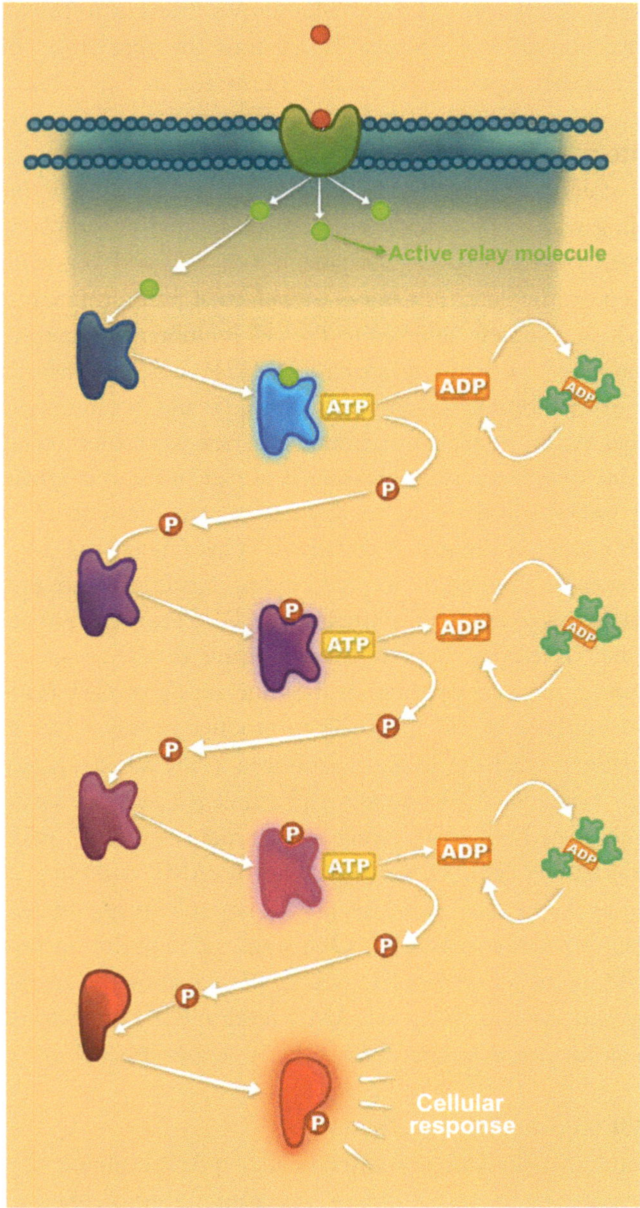

Figure 2.5 Typical signaling pathway. Signaling process indicating ligand–receptor interaction to initiate signaling, transduction of this signal through intracellular pathways and resulting cellular response. The process describes the manner in which ligand binding results in signal reception, which is then transmitted to the relay molecules. These activate another protein by phosphorylation and transmit the signal onward down the pathway. The signal may be simply transmitted or amplified along the way until a final cellular response is produced.

causing the opening of the ion channels. This allows the ions to flow in or out, resulting in the transmission of the electrical signal into the receiving dendrite. A typical ligand-gated ion channel signaling pathway is illustrated in Fig. 2.4A.

2.3.2.1.2 G protein–coupled receptor signaling pathway

GCPRs work in conjunction with a G-protein attached to a GDP molecule, which is activated by binding with GTP. These receptors are located in the plasma membrane but have a binding site on the extracellular side. Once a ligand (normally hydrophilic) is bound to the receptor, the latter undergoes a conformal transformation to attract an inactive G-protein, which is loosely attached to the cell membrane. Subsequently, the inactive G-protein sheds its GDP molecule to attach to a GTP molecule and is at once activated. The G_α portion of the trimeric G-protein with α, β, and γ subunits detaches from the GCPR and diffuses along the plasma membrane to activate a membrane enzyme, which, in turn, may activate other secondary messengers as part of the cell signaling pathway. The active membrane enzyme will return to its inactive state through the hydrolysis of the GTP back to GDP once the signal is passed on. This process can be repeated as many times a signal is received through a receptor-binding ligand but can also stop if no more stimulus is present at the receptor. The presence of GPCRs is quite common in the early stages of embryonic development as well as in many of the sensory functions such as vision. One important practical example in the functioning of GPCRs is the decomposition of glycogen to glucose by the liver cells to produce energy when stimulated by epinephrine or adrenaline, which is the signal produced in the adrenal glands located close to the kidney. In this example, the G-protein alpha unit released on activation binds to the membrane enzyme adenylyl cyclase, which then produces cAMP, which is possibly the most used messenger in all cells. The catalytic part of cAMP continues this signal through the activation of a protein kinase A, which plays a key role in glycogenesis by phosphorylating other protein kinases downstream. One single epinephrine molecule when bound as a ligand to a GPCR can cause amplification of the signal by the repeated production of protein kinase by the cyclic activity of cAMP produced by adenylyl cyclase. Fig. 2.4B depicts the mechanism corresponding to the GPCR signaling pathway.

2.3.2.1.3 Inositol triphosphate pathway

The GPCR is also involved in the activation of phospholipase C through the G_α subunit. This enzyme, in turn, will cleave a membrane lipid called inositol phospholipid (IP), resulting in the inositol molecule, which is soluble. Inositol is phosphorylated by three phosphate groups, which will convert it to IP_3, which activates calcium channels in the endoplasmic reticulum (ER) and results in the movement of Ca^{2+} down the concentration gradient from the ER to the cytosol. Calcium is another critical secondary messenger, which is ubiquitous in the cell and is responsible for a number of cell functions.

2.3.2.1.4 Receptor tyrosine kinase signaling pathway

Another cell receptor signaling also common in animal cells is receptor tyrosine kinase signaling, which is based on the action of specific enzymes. These RTKs occur in pairs with each having an extracellular ligand binding site for small, generally hydrophilic molecules. These binding sites protrude into the extracellular space and are attached to a single membrane-spanning alpha helix protein, with an intracellular region consisting of multiple tyrosine kinase proteins which serve as binding sites for phosphates. The signal ligand consisting of two molecules in the form of a dimer attaches to the pair of binding sites on the RTK receptors promoting the dimerization of the RTK pair. As a result, the intracellular tyrosine kinase regions of the RTKs are activated and cross-phosphorylate the dimerized tail sections in the cytosol. Each phosphorylated tyrosine is then bound to an adaptor protein kinase, which acquires the capability of now activating relay proteins which then transmit the signal down a specific signaling pathway to elicit a cellular response. This signal is inhibited if either the signaling dimer is somehow prevented from binding to the receptor or if tyrosine phosphatases dephosphorylate these intercellular proteins. This RTK signaling pathway is illustrated in Fig. 2.4C.

2.3.2.1.5 Mitogen-activated protein kinase pathway

An important pathway is *Ras*-activated MAPK signaling. MAPKs are a group of serine and threonine kinases normally activated by extracellular stimuli, of which, c-Jun N-terminal kinase (JNK) and extracellular signal-regulated kinase (ERK) have been well studied. ERK is responsible for *c-fos* gene activation, leading to cell growth, whereas *c-Jun* phosphorylation is induced by JNK, which is associated with response to stress stimuli and also apoptosis. The RTK receptors with the cytoplasmic enzymatic tails are activated by ligands, which are growth factors, and in this case, they are specifically called epidermal growth factor receptors. Once activated, the RTK molecules are dimerized, which further results in the activation of the intracellular tyrosine residues through cross-phosphorylation. On cross-phosphorylation, one of the phosphate-bound tyrosine residues is activated and serves as a binding site for growth factor receptor-bound protein 2 (*Grb-2*). This adaptor protein serves to recruit another protein called *Sos*, which, in turn, catalyzes the binding of the protein *Ras*, which is a small monomeric G-protein attached to the cell membrane via a lipid moiety and either bound to GDP in its inactive state or becomes active when GDP is exchanged for a GTP molecule. The *Ras* protein is extremely important for cellular proliferation and also implicated in responses to extracellular stimuli. *Ras* then activates another membrane-bound kinase enzyme called *Raf* found in the cytoplasm, which furthermore activates a series of MAPK (also called MEKs) through a phosphorylation cascade. MAPKs are activated by MAPK kinases (MAPKK), which, in turn, are activated by MAPK kinase kinases (MAPKKK). MEKs, in a final extranuclear step, activate ERKs, which are then transported into the nucleus. Inside the nucleus, the ERKs trigger gene transcription, which causes the synthesis of a variety of proteins for the growth of the cellular elements and ultimately results in cell division. A continuous functioning of this signal pathway would imply negative consequences through uncontrolled cell growth. Thus this cellular growth function ceases if one of the phosphorylation steps is terminated or when the *Raf* protein, which has an intrinsic clock to produce phosphatase, is turned off, once the required cell growth has been accomplished.

Apoptosis, or programmed cell death, is an integral part of the physiological function of a cell. An inhibition of apoptosis will result in uncontrolled cell growth, which may lead to pathologies in the human body. The reader is encouraged to look up other pathways, such as PI3-kinase signaling and the manner in which cell survival is affected when such pathways are active. An important point to keep in mind is that redundancy is the norm in cell activity so that even when one signaling pathway is nonfunctional as a result of genetic mutation, other pathways are able to replace these in carrying out the functions of the cell.

2.3.2.1.6 Intracellular signaling pathways

There is a small but important group of receptors found in animal cells, which are not located in the cell membrane but rather in the cytosol. The ligand for these receptors is a small hydrophobic molecule that is able to traverse the phospholipid bilayer and bind to the intracellular receptor, which can be either located in the cytoplasm or is able to translocate into the nucleus region for binding. A common example of this type of signaling is the function of a steroid hormone or thyroid hormone, which binds to a receptor in the cytoplasm. Subsequently, the hormone–receptor complex passes through the nuclear double membrane to activate transcription factors inside the nucleus, which ultimately results in gene expression and, consequently, protein synthesis. A typical intracellular signaling pathway is illustrated in Fig. 2.4D.

2.4 Response

Signal transduction normally has the goal of regulating the cell's activities by producing a physiological response from the cell, either in the nucleus or cytoplasm. Although the signal transduction itself may be viewed as a complex process, tangible results are obtained as a consequence of signal transduction through rearrangement of the cytoskeletal elements, change in metabolic activity, or protein synthesis through gene transcription.

Changes in cell shape are more common than is often contemplated. When muscle cells are stretched, there is a distinct alteration in their shape. The cardiac cells are constantly undergoing shape change, albeit, temporary. The cells that line the respiratory tract and blood vessels are constantly experiencing a pulsatile change in shape. The interchange of cells between lymph nodes and blood vessels is another example where cells have to "squeeze" through tight spaces, and hence, a relatively drastic change in shape may be imperative. The lengthening and shortening of cytoskeletal elements during cell division are yet another practical example of cellular response to signals. In addition, the formation of filopodia or lamellipodia, which aids in cell migration and cell-to-cell communication, is representative of conformal change in the cell.

The metabolic activity in cells is clearly driven by the signals received through specific ligand binding to receptors. Catabolic pathways in the cell break down complex molecules to produce simpler molecules, whereas anabolic pathways are involved in buildup of the complex compounds. One example that was already mentioned is the hydrolysis of glycogen to produce glucose to supply the cells with energy. Protein synthesis is yet another example of a metabolic pathway, which occurs as a result of signal transduction.

A third and very important cellular response is gene transcription, resulting in protein synthesis. This is the climactic result of most if not all of the cell signaling processes. Although we will not get into details of protein synthesis, it is important to mention that as the first step in this process, the appropriate gene is activated by a specific transcription factor or gene regulatory protein. RNA polymerase II enzyme is then used to make a copy of the gene called pre-mRNA using the coding strand of the DNA. This pre-mRNA is converted through posttranscriptional modifications to mRNA and finally results in the production of the correct protein by the ribosomes located in the rough ER. Fig. 2.6 is an illustrative flowchart of the protein synthesis process.

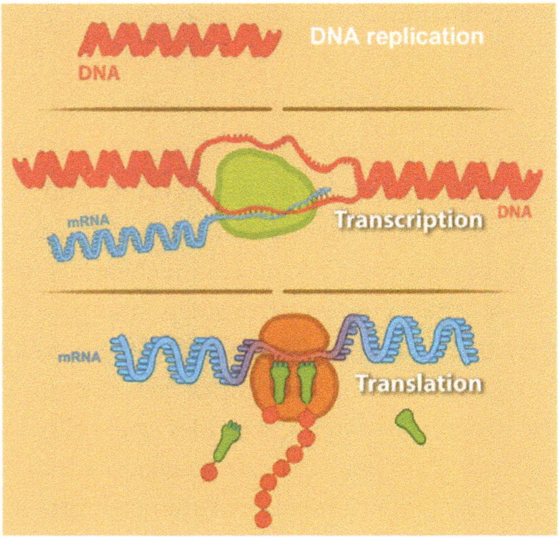

Figure 2.6 Protein synthesis. Schematic description of protein synthesis accomplished by gene transcription and translation. The double-stranded DNA is "unzipped" by RNA polymerase, which also serves as a complementary mold for transcribing the nucleotide pattern, which is exposed during this process. The mRNA strand that is produced as a result is later used in the nucleus for the required protein synthesis during translation.

2.5 Mechanical signaling in cells

There is an abundance of activity in the human body that involves the sensing of and response to mechanical signals. Shear stress and pressure, which result from the process of pumping blood and the neural signals generated as a result of pressure in hearing and touch, are simple examples. In developmental biology, separation into the left and the right sides of the fetus is attributed to mechanical stimulus. Table 1.1 indicates examples of physical functions in the human body and typical values of mechanical force magnitudes experienced by cells, cell components, and tissues in the human body. It is important to appreciate the extremely small forces that can result in a sizable magnification to produce a physiological function at the tissue or organ level. For example, a membrane tension of 0.012 N/m is estimated as necessary to stimulate stretch-activated ion channels. The amplified effect caused by these small forces in aiding Ca^{2+} transport, which is important in many large-scale physical processes must be appreciated. In addition, a traction stress of 3.0—5.5 kPa may be required to encourage migration in normal cells, whereas cancerous counterparts require a lower stress level to cause migration. This lines up very well with the idea that cancer cells have a greater tendency to migrate.

As we have learned in the introduction, the process by which cells react to an externally applied mechanical stimulus is called mechanotransduction. The process of mechanotransduction can be subdivided into three phases: (1) mechanosensing or mechanoreception, (2) signal transmission, and (3) target activation. Each stage involves a number of physical and structural elements that are critical for this process. It will be useful to gain knowledge of the structural aspects of mechanosensing proteins in the extracellular matrix (ECM) and parts of the cytoskeleton that participate in the transmission of the mechanical signal on the nucleus.

2.5.1 Cell structural elements that are important in mechanical signaling

Before delving into the actual mechanism of signaling and response to an applied mechanical stimulus, it is useful to understand the structural elements in the cell cytoplasm which participate in this process. The actual mechanisms will be dealt with in later chapters. In a general sense, the network of proteins that give shape to the cell and is involved in a wide variety of cellular functions is called the cytoskeleton. It is not correct to visualize the animal cell as solely consisting of a soft cytoplasm. An intricate and meaningful network of structural proteins provides the cell its shape. However, this structural mesh is not a rigid hardware but can disassociate and associate back again to maintain the effective cell shape. The cytoskeletal elements provide anchors to the organelles in the cell and are also thought to bind the nucleus to an intermeshing lattice of protein fibers to provide it with some sort of positional stability. Vesicular transport inside cells is also envisaged with the cytoskeleton acting as a rail guide to moving proteins to various regions in the cell.

Cell motility is another aspect, which involves certain cytoskeletal proteins whose functional roles cause the complete translation of a cell from one position to another or temporary changes in cell shape to accomplish certain functions. For example, the development of cilia and flagella directly involves the cytoskeleton. In the former case, cilia are involved in sensing as with the primary cilia and with particle movement as in cilia, which line cell surfaces. Flagella act as locomotors to cause motion of the sperm cells over relatively longer distances. Cell reproduction is yet another activity in which the cytoskeleton plays an active role, especially during the cytokinesis stage. Furthermore, the contraction of a cell during muscle movement requires the activation of the actin–myosin apparatus, which is an essential function of the cytoskeleton.

Three structural proteins that make up the cytoskeleton are briefly described.

1. Microfilaments: the smallest of the cytoskeletal elements, microfilaments are composed of two intertwined polymer chains of the protein actin, as seen in Fig. 2.7. Each chain, in turn, is made of actin monomers called G-actin for globular actin. In its basic form, actin is present in cells as a globular protein (G-actin), which is then polymerized into long actin chains, also called actin filaments. F-actin or actin filaments are about 5–7 nm in diameter and come in a variety of lengths. These chains themselves are interwoven into actin filaments (F-actin) consisting of two helical strands. One end of these actin filaments is bound to transmembranal integrins through a series of focal adhesion proteins forming the focal adhesion complex (FAC), whereas the other is intertwined with myosin to form the actin–myosin machinery and connected to the nucleus. Although actin filaments are present as linear chains, they are also capable of forming three-dimensional structures, especially in the cortex, the cytoplasmic outer layer in the vicinity of the cell membrane, which provides rigidity to maintain cell shape. Actin fibers, also called stress fibers, are very important in cell adhesion because they aid in the connection of the focal

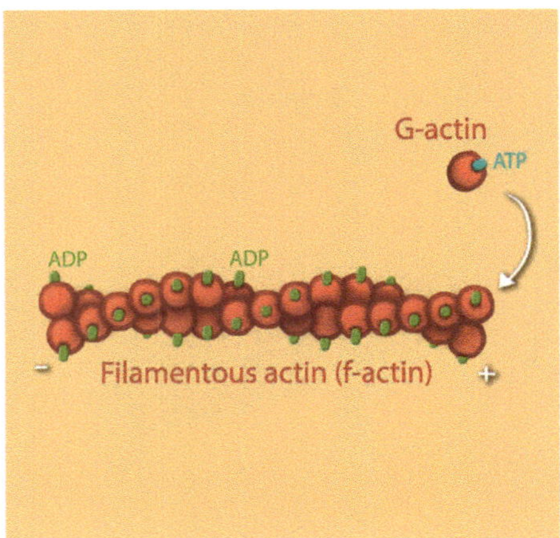

Figure 2.7 Structure of actin.

adhesions to the internal structure of the cell. They effectively maintain a network of tensile forces to maintain cell shape and help in cell motility. In the latter case, the growth and shrinkage of the actin fibers by polymerization and depolymerization, respectively, results in the appropriate development of the pseudopodia, which are essential for çell movement. Actin fibers are also instrumental in the formation of the contractile ring around the cleavage groove during the process of mitosis, whose shrinkage results in the division of the parent cell into the two daughter cells. In regions of the human body where exchange of nutrients and waste occurs across the plasma membrane, actin filament bundles make up the microvilli, which line the tissues. In muscle movement, the actin—myosin complex is active in that the myosin motors slide past the actin filaments to bring about muscle contraction at the cellular level.

Actin microfilaments are made up of G-actin globular units, which form intertwined F-actin filaments. These are very important in the transmission of stress from the ECM toward the cell interior.

2. Microtubules: This second group of elements, which make up the cytoskeleton, is the largest in size with a diameter of about 25—30 nm. Microtubules are hollow, rod-like structures constructed with tubulin protein dimers, which are formed from α- and β-tubulin units. The tubulin rods are subjected to assembly and disassembly where a relatively faster rate of association and disassociation of the tubulin molecules occurs on the "plus" end compared with the "minus" end. The hollow tubulin rods initially form as a sheet, which then rolls into a hollow cylinder shape when the appropriate 9-2 structure is attained, as shown in Fig. 2.8. Microtubules are extremely resistant to compressive forces and hence enable the cell to maintain its shape. Similar to actin filaments, these also serve as tracks to direct vesicle transport to different regions of the cell. The assembly—disassembly property of the microtubules is significant in cell division during

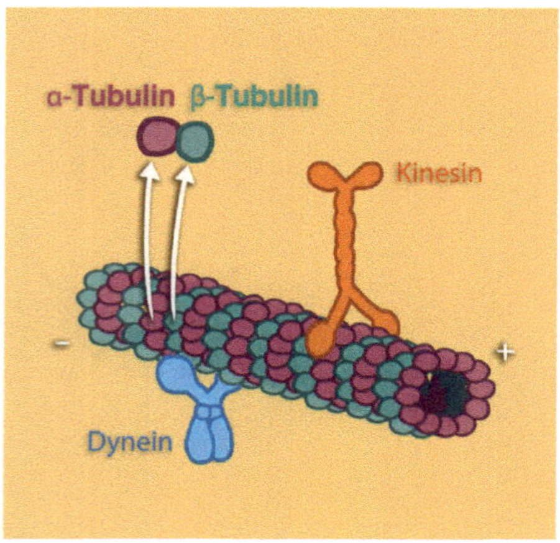

Figure 2.8 Structure of microtubules.

which one end of the microtubule attaches to centromeres of the pair of replicating chromosomes by means of kinetochores. The other end of the microtubule is anchored to the centrosome, which is commonly called the microtubule organizing center. During mitosis, the microtubule shortens its length via disassembly and hence enables the breaking away of duplicating chromosomes toward the daughter cells being formed through cell division. Flagella and cilia are other flexible appendages to the cell or cell extensions, which are constructed from microtubules to facilitate the movement of single cells.

Microtubule structure consisting of α-tubulin and β-tubulin units, which combine to form hollow microtubules, acts as conduits for the transport of kinesin and dynein motor proteins. These are also extremely important in cell division.

3. Intermediate filaments (IFs): These fibers are about 10 nm in diameter and are present in the cell with the principal role of providing skeletal support, maintaining the shape of the cell, and providing mechanical strength to tissues. The IFs are composed of two helically twined protofilaments, which are further arranged in an 8-protofilament bundle to form the IF. Six types of IF proteins are currently known, and these combine in different manners to form the IFs, which are cell specific. For example, in epithelial cells, they are expressed as keratin, whereas in glial cells, they are expressed as glial fibrillary acid protein. In addition, neurons express neurofilaments, and vimentin is found in endothelial cells. IFs are more stable than the other filaments and hence provide support to the nucleus by a network made up of laminin in the nuclear envelope. These are an important component in desmosomes and play a key role in cell—cell adhesion and in attachment to the ECM through cellular structures called hemidesmosomes.

In addition to the main structural cytoskeletal proteins mentioned earlier, some other proteins, which participate actively in the process of mechanotransduction, are briefly highlighted below:

1. Integrins: One of the most important cell adhesion molecules is integrin, which is actually a dimer consisting of two subunits, namely an α component and a β component. This dimer is a transmembrane protein with the α and β components connecting with either a cell adhesion molecule in another cell or with the ECM in the extracellular domain. The transmembranal section is made up of an alpha helix, which then attached to the intracellular domain connected to a complex group of proteins, in turn, bonded to the actin filaments. Integrins can also bind to other integrins in neighboring cells and play a key role in cell—cell adhesion. Briefly, it is important to note that extracellular mechanical stimulus is transmitted via the integrins into the cells through the actin filaments or stress fibers. The structure of integrins and their role in mechanotransduction will become clearer in later chapters.
2. Cadherins: They are another group of cell adhesion molecules. The cadherin superfamily has more than 150 cadherin proteins. These are mostly integral membrane

proteins, although there are exceptions. T-cadherin, for example, is not a transmembranal protein and binds to the phospholipid bilayer through a lipid moiety. All cadherins have a special domain in the extracellular space called the extracellular cadherin domain. There are mainly five families of cadherin proteins, of which Type I cadherins are the ones involved in cell—cell interactions. These form structures called adherens junctions, and such cell—cell binding is highly important for maintaining the extracellular calcium ion content. Similar to integrins, adherens junctions are more clearly illustrated in a later chapter.

2.5.2 Mechanosensing or mechanoreception

Mechanosensing may be described as the ability of a cell or group of cells to sense mechanical cues from the microenvironment on the application of load/deformation or simply sense changes in the stiffness, roughness, topography, or orientation of the substrate on which they grow. This is the initial phase in the mechanotransduction process in which the cell receives or detects the external mechanical stimulus and relays it into the cell. Thus a mechanosensor is any simple or complex biological element, which is capable of sensing changes in terms of the mechanical stimulus. This entails that the mechanosensor is in direct contact with the ECM environment, which is the first interface with the mechanical stimulus and is capable of detecting changes in load, deformation, pressure, or fluid flow on the plasma membrane. Possible entities include the primary cilium, which has been singled out to sense changes in fluid shear at the plasma membrane surface or protein structures in the ECM, which are somehow connected to intracellular cytoskeletal elements. Molecular mechanisms by which cells can perceive their mechanical environment and respond to such stimuli have only been recently investigated. Several transmembrane receptors are critical for mechanoreception. These receptors, which include integrins and cadherins, are responsible for receiving the signals resulting from applied mechanical stimuli. Integrins connect fibronectin fibers from the ECM to FACs in the cell cytoplasm, which are further connected to the cytoskeletal elements inside the cell. Cadherin proteins also play a role in the reception of mechanical signals because they bind adjacent cells and transmit signals received from cells to neighboring cells.

2.5.3 Signal transmission

The mechanical signal received by the cell is transmitted to the appropriate targets inside the cell through biochemical pathways or simply relayed through the cytoskeleton. Various signaling pathways have been reported to be activated after mechanical signal reception. Src, RAS, MAPK, and other pathways may be activated depending on the type of mechanical stimulus received—whether it be tension/stretch, compression/contraction, or shear/distortion. In addition, the type of cell that receives the signal the intensity and periodicity of the signal itself may influence the type of signal that is transduced. Mechanical signals may also be transmitted to the nucleus through the cytoskeletal network and may directly affect the synthesis of DNA and, as a result, accelerate transcription processes and even play a direct role in mitosis.

2.5.4 Target activation

Proteins are normally activated as the signal reaches specific targets and ultimately alter cell behavior or physiology through various molecular mechanisms. Mechanical signals have been speculated to directly influence stretch-activated ion channels or possibly activate biochemical signaling pathways. It is also possible that the mechanical forces may expose regions in folded protein structures that were closed and inactivated before stretching. Increased protein synthesis, accelerated cell migration, and well-defined cell orientation patterns have been observed as a result of mechanical stimulus. However, the unlimited increase in force intensity on cells has also been reported to cause apoptosis or abrupt termination of gene expression.

2.5.5 Gene expression

The process of synthesis of a cellular protein is generally called gene expression. Protein synthesis is a complex process that is accomplished when the cell receives a signal to produce a certain protein. This process, which initiates inside the nucleus, is called transcription and consists of three steps: initiation, elongation, and termination. A promoter stimulates RNA polymerase to initiate the transcription process by unzipping the DNA strands at a specific location. The RNA polymerase also serves as a complementary mold to transcript the sequence of nucleotides from the DNA strand or gene that has been selectively exposed. A terminator ends the transfer of this information from the DNA template to the mRNA, thus accomplishing the process of transcription. The product resulting from transcription is a strand of immature mRNA, which will be used later in the translation process. In this maturation process, the mRNA produced inside the nucleus is appropriately spliced and is released to the rough endoplasmic reticulum (RER) through the pores in the nuclear membrane.

In the second step of this process of gene expression, the mRNA transcript is translated, resulting in protein synthesis. Translation occurs in the ribosomes located in the RER and is activated by the START codon. Protein synthesis continues in tune with the codons present in the mRNA strand as processed by the ribosome. When the template obtained from the DNA during transcription has been translated, a stop codon is activated, and the process of protein synthesis is completed. The protein is present in the form of a polypeptide, which can be transferred to the correct location in the cytoplasm by vesicular transport via the Golgi apparatus and may be later transferred out from the cell through exocytosis. This process was described briefly in an earlier illustration in Fig. 2.6.

In the case of transduction of mechanical stimuli, two paths are emphasized—either the applied mechanical stimulus through force or stretch/strain causes a direct change in the physiology of the cell; or the mechanical signal is transduced to a biochemical signal, which is then transmitted through various signaling pathways. The mechanical stimulus may directly effect conformational changes in the cell membrane or cause changes in the cytoskeleton through integrin proteins. On the other hand, the transduction of the mechanical signal to a biochemical signal involves receptor activation at the membrane surface followed by signaling in the cytoplasm. Shear stresses, in particular, control the action of many signaling molecules, including intracellular Ca^{2+}, G-proteins, protein kinase C, IP_3, and so on.

One interesting aspect of this mechanochemical transduction is the multiplicity in the signaling pathways that may be activated by a single applied mechanical stimulus, in stark contrast to specific signal activation in the case of chemical ligand molecules. Thus instead of inducing specific conformational changes to the receptor, mechanical signaling may indeed activate several membrane proteins. Shear stresses, in particular, activate potassium ion channels in the upregulation of NOS while also modulating G-proteins and protein kinase C in platelet-derived growth factor expression. Applied shear stress on endothelial cells alters lipid fluidity and activation of G-proteins. Thus molecular transport inside the phospholipid cell membrane as a result of applied shear stress becomes important in the initial phase of mechanotransduction. Similarly, the reorganization in integrin clustering will also regulate the timing of the signal produced as a result of mechanical forces.

In the cell signaling process, transcription factors are activated through phosphorylation of various proteins. These translocate to the nucleus and act on different genes. These transcriptional factors act on their targets over the course of a variety of periods based on their functional roles. As a result, multiple genes are activated because of an applied mechanical stimulus. Gene expression itself is characterized by temporal and spatial responses, which in the overall scheme of homeostasis is critically important for the proper functioning of all tissues and organs as well as serving as a deterrent to pathogenesis and disease progression. The overall responses of cell to applied mechanical stimuli share common signaling pathways with chemical signaling responses. Ultimately, only a few fundamental mechanisms are used in response to various stimuli. The complex interactions between these differing signaling pathways merit a careful study and understanding to address the multiple problems in human health and the provision of viable solutions to deal with disease.

In the following chapters, we will focus on specific cell types and their response to mechanical stimuli. These are grouped into some of the relevant physiological systems in the human body to enable the reader to clearly understand the diverse mechanisms of mechanotransduction present.

Review and Project

1. What are the different cell communication methods? Describe one relevant physiological process corresponding to each cell communication method.
2. Enumerate a few cell signaling pathways. Research other signaling pathways and present these graphically for deeper understanding.
3. What are the three steps in mechanotransduction? Describe three physiological processes where this process is observed in the human body, focusing on the three phases of mechanotransduction.

CHAPTER 3

Cell signaling under mechanical stimulus in specific cell types

3.1 Introduction

In Chapter 2, Cell Signaling: An Overview, the emphasis was placed on the general mechanism of cell signaling, including reception, transduction, and response, in some detail. In this chapter, before we turn our attention to cell response to mechanical stimulus in individual tissues, we will delve into the details related to the signaling machinery in cells as well as some general but fundamental concepts within the realm of cell signaling. Later on, in this chapter, the focus will be on various signaling pathways in specific cell types when subjected to a mechanical stimulus. Most, if not all, cells are subjected to mechanical stimulus in the human body at the organ or tissue level. This stimulus is critical to homeostasis and modeling and remodeling of different tissues, the absence of which leads to many pathologies. It must be remembered that a wide gamut of cells responds with varied physiological responses depending on the type of mechanical stimulus to which each is subjected. Although the general mechanism in various cell types involves the three essential steps mentioned earlier and discussed in fair detail, each specific cell type has a particular signaling pathway that is distinct from other cell types based on the type of mechanical stimulus. This does not imply that there is no overlap among the cell signaling pathways in the different cell types. Even though the common initiator is mechanical stimulus in the process of mechanotransduction, the final response is specific to each cell type, although some of these responses can be grouped under certain broad categories such as multiplication and differentiation. In addition, the cells in each tissue type are complex and possess characteristics and roles different enough to exhibit varied responses under the same mechanical stimulus. Cells will also recognize the distinct signals coming from themselves and from their environment and tailor their responses according to physiological priority.

3.2 Forces acting on cells

Fig. 3.1 pictorially demonstrates the types of mechanical stimuli at the cellular or subcellular levels: (A) tension, where the cell is stretched in the direction of the stimulus; (B) compression, where cell dimensions are decreased in the direction of the force; (C) shear, where the mechanical stimulus is applied parallel to the cell surface causing distortion; (D)

Mechanotransduction
DOI: https://doi.org/10.1016/B978-0-12-817882-9.00005-7

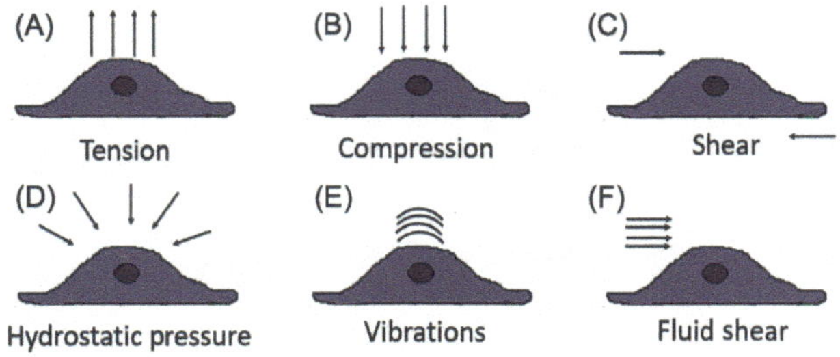

Figure 3.1 Types of mechanical stimuli. Types of mechanical stimuli that can be experienced by cells, including (A) tension, (B) compression, (C) shear, (D) hydrostatic pressure, (E), vibration, and (F) fluid shear.

hydrostatic pressure, where force is exerted uniformly and causes a decrease in cell volume; (E) vibration, where an oscillatory stimulus is applied; and (F) fluid shear, where the apical side of the cell membrane is subjected to a parallel force by fluid flow.

The specific force to which a cell is subjected depends on the environment in which it resides. However, it is rather tricky to define the exact force that acts on a cell as a result of an external mechanical stimulus applied to the tissue and transmitted onward to the cell. For example, cells in long bones, such as the femur and tibia, are bent under the action of axial forces to include both tensile and compressive stresses depending on the bending arc. These also experience fluid shear stress based on the direction of fluid flow in the channels based on the bending direction. Similarly, cells in tendons, although stretched, may experience a compressive force from the collagen fibrils surrounding the cell. Thus the response of such cells to this multiforce scenario may be quite complex. In general, the force magnitude on the cells is often in the order of a few piconewtons. Other than the type of stimulus, the duration of stimulus application and the pulsatile or oscillatory nature of the signal are important characteristics to be kept in mind.

3.3 Modes of mechanosensing

In the following section, some general concepts already mentioned in Chapter 2, Cell Signaling: An Overview, will be explained in more detail. Although there are several pathways for the detection of mechanical signals, stretch-activated ion channels and integrins are possibly the most important. In fact, ion channels may be critical in most if not all initial responses to a mechanical stimulus. Calcium ion channels on activation may cause drastic changes in Ca^{2+} concentration both extracellularly and intracellularly within a few milliseconds. Integrins may be activated through the stimulus applied to components of the extracellular matrix (ECM). Fig. 3.2 depicts a general model for mechanosensory detection, transduction, and response, which divides the overall mechanotransduction process in various steps, indicating a sequential operation of these phases.

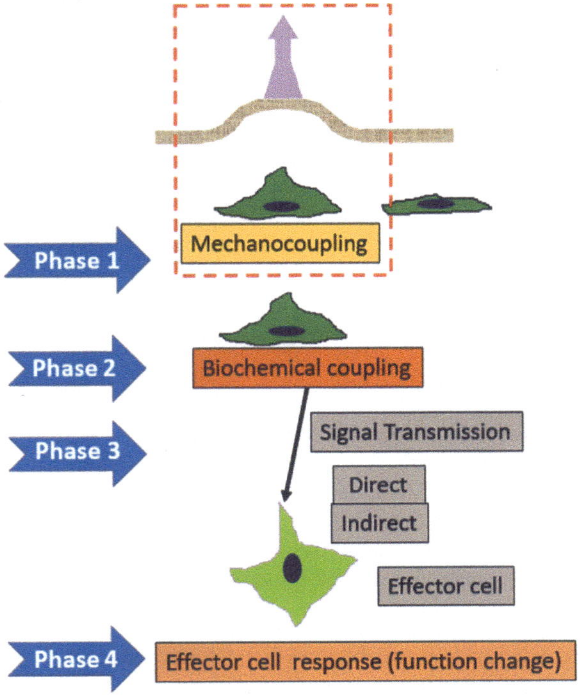

Figure 3.2 Phases of mechanotransduction. Simple illustration of a general model for mechanotransduction showing the application of mechanical stimulus (Phase 1), mechanosensing and signal transduction (Phases 2 and 3), and cell response (Phase 4). A variety of stimuli and the corresponding operation of appropriate mechanosensors is possible. Signal transduction may be quite complex, and the resulting cellular response is unique depending on the cell type on which the stimulus is applied.

The three distinct stages in mechanotransduction, including sensing/reception, transduction, and response, are clearly marked in Fig. 3.2. In general, the mechanical stimulus is either sensed directly by the cell membrane that then activates the stretch ion channels or received through the protein tyrosine kinase receptors. A rapid change in the calcium concentration in the cell and its environment has been experimentally demonstrated on the action of mechanical stimuli on the plasma membrane. Ion channel gating is a result of deformation-induced changes in shape in the biphospholipid membrane. Therefore it is hypothesized that these deformational changes may be responsible for inositol trisphosphate (IP_3)-activated calcium channels. Although proof has been provided for the necessity for IP_3 to activate calcium channels, there is some doubt about this mechanism because the protein heparin, that binds to the IP_3 receptor and blocks the propagation of the calcium wave, also appears to be present. Alternately, IP_3 also may be transported through the gap junctions (GJs), and hence, it is believed that GJs may be the fastest and most important conduit to convey the primary mechanical signal to neighboring cells.

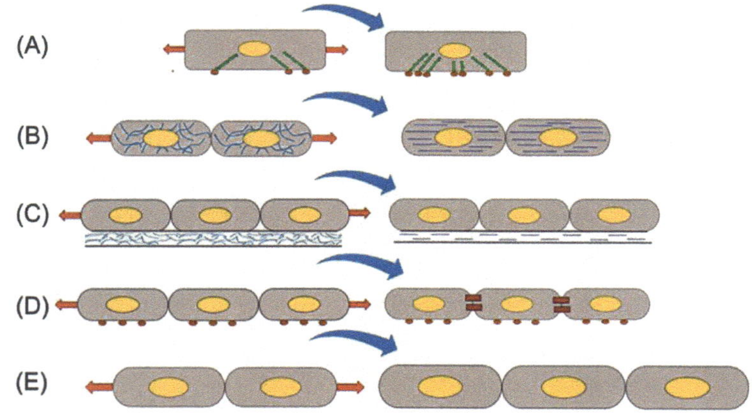

Figure 3.3 Possible cell responses based on mechanical stimulus. A variety of cell responses are possible as a result of mechanotransduction as shown. (A) Strengthening of anchoring proteins by the formation of additional FA clusters and their attachment to the actin stress fibers. (B) Rearrangement of the cytoskeletal elements, which align in the direction of the applied stimulus. (C) Remodeling of the extracellular matrix. (D) Relocation of proteins to other regions of the cell. (E) Cell division encouraged by the applied stimulus. *FA*, Focal adhesion.

The alternate mechanism is perturbance of the ECM, which then transmits the signal to the cytoskeleton via the transmembranal integrin receptors. These integrins bind intracellularly through the β subunit to α-actinin or talin, whereas the extracellular connection is made to the matrix molecules, including laminin, fibronectin, or even collagen fibers. It is now known that all the integrin present in the cell is not used in forming focal adhesions but is used on a need basis for the association and dissociation of the focal adhesions. Focal adhesions themselves are complex clusters of at least 40–50 protein molecules and normally take various days to nucleate and mature. However, these are known to dissociate and reform fairly quickly under the action of a mechanical stimulus to cause a conformal change and spreading of the cell or induce motility in the cell. The kinetics of focal adhesion nucleation, growth and maturation, their density, and distribution under the action of an external stimulus is unclear at present. Similarly, the redistribution of these focal adhesions during cell migration and apoptosis is poorly understood. During apoptosis, the focal adhesions become inactive, allowing for a gradual decrease in cell spreading and instead encourage rounding of the cell before programmed cell death. Some interesting illustrations of cellular responses to applied mechanical stimulus are given in Fig. 3.3.

A number of mechanosensing structures are available to the cell to sense and respond to applied mechanical stimuli. Focal adhesions, intercellular adhesion molecules (ICAMs), and other signal receptors have already been mentioned in earlier sections as the first front in the reception of mechanical signals. The second-tier transmitters of these signals include the cytoskeletal components such as actin filaments and microtubules. Another group of these cytoskeletal elements are intermediate filaments, which are more stable than the actin fibers. Thus mechanical forces either increase the tension in these structures or produce conformal changes in the cell membrane proteins. As can be appreciated in Fig. 3.3, a number of cell responses are possible on the application of a mechanical stimulus. For example, intermediate filaments are phosphorylated under the influence of tension forces by kinases and recruited to cell adhesion molecules. In addition, the ECM is remodeled by the reorganization of collagen IV fibrils under the action of an externally applied mechanical stimulus. Also, proteins from one part of the cell move and organize in different regions as per the needs of the cell function, as in the case of vinculin from focal adhesion, which moves under the action of a mechanical stimulus and binds to the adherens junctions to strengthen cell—cell adhesion. Cell division and cell orientation are other physiological functions affected by mechanical signals. In general, these are the types of primary cell responses to forces acting on a cell or groups of cells. Other transcriptional activities such as β-catenin translocation to the nucleus under compressive stresses in epidermal cells are more of a secondary response.

3.4 Factors affecting general cell response

Again, in this chapter, the emphasis is laid on the mechanosensory machinery that is available in cells, in general, to respond to mechanical stimuli. As mentioned briefly earlier, cells are able to detect and respond not only to direct mechanical stimuli but also to perturbations in their ECM. The normal outside-in signaling involves integrins being the first responders, after which this signal is transmitted to the cytoskeleton through mechanosensory machinery, including the focal adhesion molecular complex involving focal adhesion kinase (FAK), Src, paxillin, tensin, talin, and vinculin up to the nuclear envelope. This, in turn, elicits one or various signaling cascades as a result of conformal changes in proteins and regulation of protein kinase activity. The end result is transcriptional either through gene expression or a conformal change in the cytoskeletal structure from an inside-out signaling.

Although it is not entirely accurate to correlate the behavior of cells cultured on an artificial substrate to their actual behavior in the human body, experimental data with two-dimensional cell cultures provide us with critical information with respect to their behavior in actual tissues. Three factors are reported to influence how cells respond to an applied mechanical stimulus: (1) cell contractility in response to the applied stimulus through rearrangement of the cytoskeletal framework; (2) characteristics of the substrate, including rigidity, surface roughness, texture, and dimensionality; and (3) the temporal and spatial characteristics of the applied stimulus.

3.4.1 Stress state of the cell

Other than an externally applied mechanical stimulus, cells assume a "residual stress state" as soon as they adhere to a substrate. It is now generally assumed that cells occupy a low

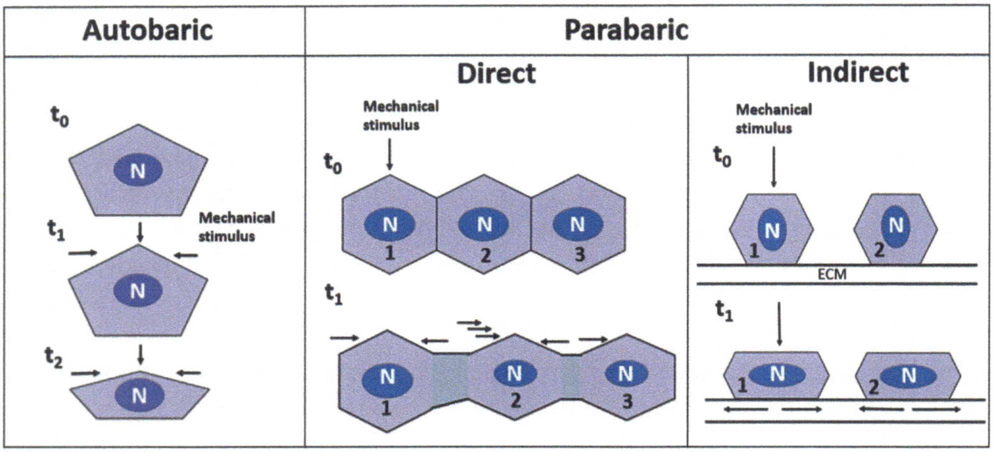

Figure 3.4 Internal stress equilibrium in cells. (A) Autobaric effect: Single cell indicating change in its cytoskeletal structure and form and hence its stress state as a function of time (t_0, t_1, and t_2) after receiving a mechanical stimulus. (B) Direct parabaric effect when cell, which is the recipient of the mechanical stimulus, transmits the signal to its neighbors directly. (C) Indirect parabaric effect when the receptor cell undergoes changes in its cytoskeletal structure and transmits force to the adjacent cell via the substrate (ECM). *ECM*, Extracellular matrix.

energy configuration, which involves a basal tensegrity state in which they adhere to the substrate by cell spreading and positioning. This stable or equilibrium configuration takes into account a residual stress state in which a cell exists. This is often superposed by self-loading, where the cell "loads" itself by contracting, referred to as the autobaric state, that is, a cell imposes a stress or mechanical stimulus on itself. This self-imposed autobaric stress is deemed to be responsible for cell motility and cell division as part of its physiological functions even in the absence of an externally applied mechanical stimulus. Cells, subjected to a cyclic stress, polymerize actin filaments, which then aggregate to support the applied external load and transform to a newer stress state via an autobaric effect. On the other hand, when a cell applies or transmits an incoming mechanical signal to neighboring or adjacent cells either directly through cell—cell adhesion molecules as in adherens junctions or by transduction through indirect means without actual contact, the process is referred to as parabaric effect. This mechanical stimulus can be transferred from the cell, which receives the signal and transmits it indirectly to its immediate neighbors via the substrate or ECM. Fig. 3.4 provides a graphical explanation of the autobaric and parabaric effects as a function of time.

3.4.2 Substrate characteristics

Over the last two decades, the role of the substrate in affecting cell response had been demonstrated very convincingly. The stiffness of the substrate measured by its elastic modulus or just elasticity has a strong influence on cell function. This phenomenon is termed durotaxis or mechanotaxis, where the stiffness of the underlying substrate on which the cells are cultured or seeded affects the amount of cell spreading and the choice of position and adhesion of the cell on the substrate. It is now accepted that cells tend to migrate to regions of higher stiffness for greater stability. Another very important factor is the topography of the substrate. Cells appear to favor a certain optimum roughness level for adhesion or a certain alignment based on a preference for adhesion sites or channels or specific patterns. This phenomenon is called haptotaxis.

3.4.3 Temporal response of cells to applied mechanical stimulus

Another important aspect of cell signaling is the temporal response to the applied mechanical stimulus. Cell receptors can be activated in a matter of seconds as in the activation of some of the stretch ion channels, which can occur in a duration of milliseconds, whereas conformal changes involving the rearrangement of the cytoskeleton may take from minutes to hours. In contrast, behavioral changes may actually take days, including migration, mitosis, or apoptosis. Fig. 3.5 is a neat pictorial representation of the temporal nature of sensing, transmission, and transduction of the mechanical stimulus on a gross scale without including the details involved in each phase.

In a more specific sense, the cellular and molecular responses obtained as a result of mechanical stimulus in various mechanotransduction phases are succinctly illustrated in Fig. 3.6.

As seen in Fig. 3.6, in the time frame of a few milliseconds to seconds, signal reception through stretch activation of ion channels occurs to allow passage of ions to activate the signaling pathways. Within this extremely short period, G protein signaling is also initiated. Further along the time scale over minutes to hours, transcription and translation occurs, so also, cytoskeletal rearrangement as a result of polymerization and depolymerization processes. Over the longer period of days to weeks, cell scale

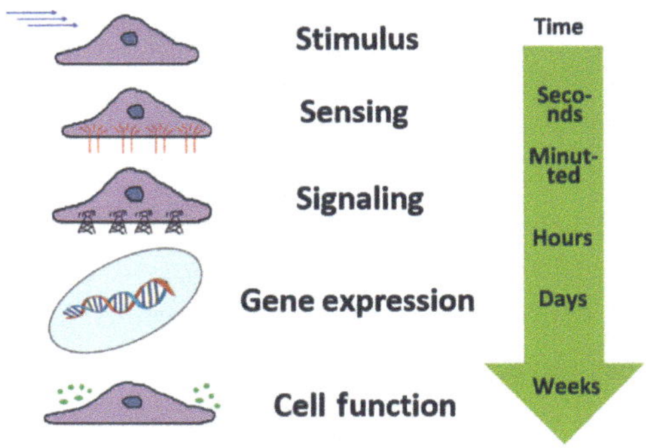

Figure 3.5 Duration of the process of mechanotransduction. Illustration of the time-lapses involved in each phase of the mechanotransduction process. Mechanosensing may take a few seconds to minutes, signaling may extend for hours, whereas the cell response could take from days to weeks.

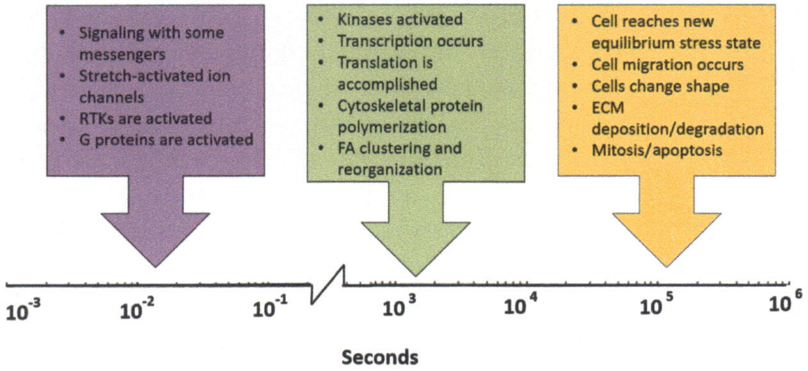

Figure 3.6 Intracellular responses to mechanical stimuli. Compendium of some of the responses in cells at the cell level and molecular level when subjected to mechanotransduction. Within seconds, intracellular signaling takes place along with the activation of stretch ion channels. G proteins are also activated. Transcription and translation activities are accomplished and some cell responses such as the polymerization of actin filaments and FA clustering and rearrangement occur from minutes to hours. Cell migration, ECM remodeling, and apoptosis are more long-term responses. *ECM*, Extracellular matrix, *RTK*, receptor tyrosine kinases.

changes including cell migration and clustering and reorganization of focal adhesions, ECM remodeling, cell division, and programmed cell death are observed.

Although more emphasis will be placed on changes in cell characteristics as a result of mechanical stimulus in Chapter 4, Cell Characteristics Resulting From Mechanical Stimulus, suffice to say that a general approach to the specific types of cells in four different tissues in the human body and the sensing and transduction of the mechanical stimulus in these cells with a focus on the associated signaling mechanisms will be provided in this chapter. Of course, the tissue most studied is bone tissue, and there is ample evidence in literature possibly because of the relative ease of testing this tissue as well as using computer modeling as a useful tool in such studies. However, pulmonary, cardiovascular, and nerve tissues have become more and more important as the scientific community understands the prevalence and relevance of mechanical stimuli in embryogenesis and development of these and other tissues.

In another characteristic of cell response to external stimuli, the resulting cell behavior is selective depending on the type of signal received. Based on whether the input signal is a static or cyclic stretch, cells respond differently in terms of growth expression. For example, the same stress fibers align in the direction of the applied load *in vitro* under static stretch while they are oriented perpendicular to the cyclic stretch with maximum efficiency at a frequency of 1 Hz. In addition, if several forces act on the same cell, each elicits a specific cellular response. During embryogenesis, valve precursor cells experience both shear stress and reversing blood flow. The latter is influential in the development of the valve structures, whereas the shear stresses are sensed by the primary cilia in valve cells.

Early studies on mechanotransduction focused on sensory cells, which are designed to produce a biochemical response when a specific mechanical stimulus is applied to them. With an increase in the understanding of this phenomenon, it has become evident that mechanotransduction is critical to the proper functioning of various tissues subjected to constant mechanical loads, such as bone, cartilage, muscle, and vasculature. Consequently, a variety of cells such as osteocytes, vascular endothelial cells (ECs) and smooth muscle cells (SMCs), myocytes, and chondrocytes all have become subjects of study of the effects of mechanical stimulus on these cells. In the following sections, the focus will be on cell signaling in specific cell types. Although some of the signaling paths may be slightly different, many of the receptors fall in the same class in many of these cell types. Nevertheless, it is worthwhile understanding the mechanisms of cell signaling operating in different cell types.

3.5 Cell mechanotransduction in musculoskeletal tissue

3.5.1 Types of cells in bone tissue

Bone cells sense and respond to a variety of mechanical stimuli. These cells can be divided into three types: (1) osteoblasts, which are bone-forming cells derived from mesenchymal stem cells (MSCs) and involved in the anabolic bone building process; (2) osteocytes, which are mature or terminally differentiated in this mesenchymal lineage, which surround themselves as the tissue matures into the osteoid and are responsible for bone homeostasis; and (3) osteoclasts, which are the bone-resorbing cells of hematopoietic origin. These three bone cell types are illustrated in Fig. 3.7.

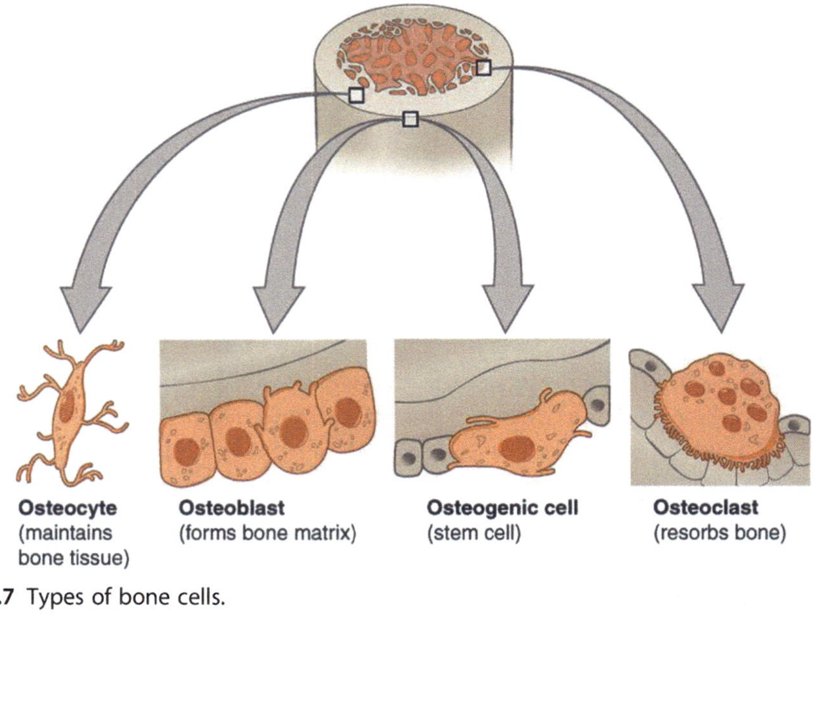

Figure 3.7 Types of bone cells.

Bone cells found in the skeletal system, including osteogenic cells, which are the progenitors of osteoblasts, which later mature into osteocytes. Osteoclasts are hematopoietic bone resorption cells. Osteoblasts are responsible for the bone modeling process, whereas osteocytes are found in mature bone tissue (CC BY 4.0).

Osteoprogenitor cells are sometimes referred to as a fourth class, although osteoblasts and osteocytes are derived from these MSCs. Mechanical stimuli have independent and different effects on each of these cell types, but their interaction can serve as regulatory factors, which are indeed critical in their proliferation and differentiation. Mechanically sensitive environments appear to have a strong effect on the osteoprogenitor cell lineage in the embryonic development stage up to osteoblast formation and final maturation into osteocyte. Early osteoprogenitors appear to respond to the mechanically prone bone marrow environment by not only altering differentiated cell types but also increasing rates of proliferation and differentiation.

3.5.2 The mechanosensory environment of bone cells

The environment to which bone cells, in general, are exposed to includes stress, strain, fluid flow, pressure, and streaming potentials. The complex interaction among these many mechanical stimuli is responsible for the overall homeostasis in the bone tissue. Based on the type of applied mechanical stimulus, bone tissue organizes its structure and mass to provide maximum strength/mass ratio through functional adaptation. The osteocyte network may eventually control functional adaptation in bone tissue by regulating remodeling and modeling as a result of mechanical stimulus. The osteocyte is conferred with the responsibility of mechanosensing in bone tissue and appears to be most responsive to fluid flow. Its unique position in the bone matrix and its long, slender osteocytic processes enable it to function as an efficient mechanosensor. It is the mechanosensory cell in bone tissue and ultimately controls both anabolic and catabolic processes in this tissue through the release of signaling molecules. The osteocyte is also responsible for maintaining equilibrium between the bone building function of the osteoblasts and the resorption role of the osteoclasts. The osteocyte is directly connected to the canaliculi through its processes and senses the alterations in the flow of blood through vessels as a result of the mechanical forces and strains applied to the bone. The motion of interstitial fluid in contact with the osteocytes in the extensive canaliculi network in bone serves to generate shear stresses and amplify the strains in the bone tissue. The differential flow of blood appears to modulate the shear stress, which acts as the stimulus that activates the bone cells. The bulk fluid flow is driven by externally applied mechanical loads rather than the pressure differential caused by the interstitial fluid flow. Many recent studies have clearly attributed the mechanosensing capacity of the osteocyte to the presence of primary cilia, which are believed to be the direct sensor of stress in osteocytes.

3.5.3 The mechanoreceptors in bone cells

Although it is universally accepted that osteocytes are the mechanosensors in bone tissue, how these cells receive the signal is still unclear. As mentioned briefly earlier, primary cilia are prime candidates as mechanoreceptors if the stimulus is transmitted via fluid flow or pressure. Conformal transmembranal proteins, which straddle the plasma membrane and make the extracellular—intracellular connection, are other candidate mechanoreceptors. A number of mechanoreceptors operate in bone cells, and these are enumerated as follows.

1. *Primary cilia* are microtubular mechanosensing structures in bone cells. Prostaglandin signaling induced by fluid flow and independent of the influx of Ca^{2+} is ascribed to be regulated by primary cilia. The regulatory role of primary cilia in suppressing osteogenic genes in MSCs is also reported. Primary cilia also modulate Wnt signaling, which can affect gene transcription, regulate the cytoskeleton, and control Ca^{2+} concentration. As more information becomes available, the role of this mechanosensor in modulating cell behavior with regard to its environment is being understood in depth.
2. *ECM-based structures*, which include *integrins*, are another group of mechanosensors, which are essential to cell attachment and resultant behavior resulting from such tethering. Various cell signals, which are sensed in the ECM because of cell adhesion, are transmitted intracellularly by these biological structures. For example, integrins have been postulated to sense and transmit signals based on fluid shear in osteocytes into the cell machinery. The spatial arrangement of integrins and their unique structure enables them to be effective mechanosensors. The role of integrin appears to be even wider, as these are triggered by both mechanical forces and fluid forces. Integrin manipulation triggers the activation of RhoA signaling molecules, which, in turn, gives rise to the formation of other focal adhesions to enable clustering of these structures, whereas β_1 integrin expression is upregulated in osteoblasts exposed to shear stress resulting from fluid flow. This, in turn, activates the specific integrin $\alpha_v\beta_3$, which colocalizes with the nonreceptor protein kinase and proto-oncogene Src. β_1 integrin activation of Src-like kinase is influenced by caveolae, and the mechanical properties of bone tissue are regulated by caveolin-1, which prevents the entry of osteoprogenitor cells and possibly inhibits β-catenin availability. Osteoblasts also activate another integrin protein $\alpha_5\beta_3$ on mechanical stimulation while geometrically modifying other linker proteins in the focal adhesion complexes (FACs) such as talin and paxillin inside the cell by applied signals in the ECM. This provides new binding sites for activation of other proteins or function as new signal initiating sites as in the case of FAKs, which are responsible for the clustering of integrins and enhanced attachment of focal adhesions to the substrate. Integrins are also reported to be important in cell membrane deformation in osteocytes by regulating interstitial fluid flow in the lacunocanalicular system.
3. *FACs* are macromolecular moieties, which are connecting structures between the ECM and intracellular cytoskeletal elements, such as actin stress fibers. The latter are extremely important in cell division and motility as well as in molecular transport of

intracellular proteins. The complex geometry of these focal adhesions invests them with the capacity to be excellent mechanosensors and further transduce the signal.

4. *Plasma membrane* consists of many lipids, lipid rafts, and anchoring and transmembrane proteins, which play important roles in the signaling system. The arrangement and abundance of these types of mechanosensors become important not only for reception but also to effectively regulate cross-talk and transmission of signals into intracellular space. The presence of caveolae is another factor that underscores the mechanosensory role played by the plasma membrane. Lipid rafts are key for the activation of extracellular signal-regulated kinases (ERK1/2) through mechanical stimulation and the expression of *c-fos* in osteoblasts.
5. *Cell–cell connections* including *cadherins* are another group of mechanosensory proteins, which are transmembranal anchoring proteins with long, extracellular domains, that connect neighboring cells to act in unison as sensor circuits. Cadherin protein complexes consist of α- and β-catenins, which are connected to the cytoskeleton through vinculin. β-catenins associate with cadherins in osteoblasts on the inner leaflet of the cell membrane. β-catenin binding to N-cadherin appears to be downregulated by fluid shear stress, resulting in nuclear translocation of this protein, as it accumulates in the cytoplasm. Another RTK protein, ephrin, appears to be necessary for the process of differentiation in osteoblasts and for cell–cell communication during remodeling. In addition, incoming mechanical signals appear to be modulated by ephrin and tuned by cell–cell connections among themselves and to the cytoskeleton.
6. *Ion channel* and *connexins* contribute as mechanosensors in bone cells. Alignment of connexin protein complexes results in the formation of GJs, which are channels for transport of small molecules such as calcium, inositol phosphates, adenine triphosphate (ATP), etc. Connexins also serve as channels for the release of prostaglandins in response to fluid shear stress. Ion channels are critical to the maintenance of an adequate amount of electrochemical gradient in cells and across the cell membrane. Mechanical strain on the cell membrane results in the activation of voltage-sensitive calcium channels, which, in turn, regulate the release of anabolic signals in response to mechanical stimuli in osteoblasts.

3.5.4 Signaling pathways in bone cells

After mechanoreception, the signal is transducted by other proteins through various signaling pathways. The prime mechanotransducers appear to be transmembranal receptors such as CD44 and also integrins, which act to convert the mechanical stimulus into the intracellular signal. Although it is unclear at present time, well-known intracellular signaling pathways such as IP_3-, Ca^{2+}-, and cAMP-dependent pathways may be ultimately responsible for signal transduction. The signal transduction results in the secretion of molecules such as nitric oxide (NO) and prostaglandins. The latter play a significant role in bone resorption because the upregulation of PGE2 as a result of mechanical stress has been shown to affect the subsequent release of cyclooxygenase-2 (COX-2). This points toward a continued release of prostaglandin (PGE2) well after the mechanical stimulus is removed, suggesting the memory retention capability of bone tissue. NO, on the other hand, is an important mediator in bone tissue response to stress and inhibits bone resorption by osteoclasts.

Once the cell senses the mechanical stimulus through the mechanosensor, the signal is now transduced intracellularly downstream to cause changes in cell shape, motility, or gene expression. The signal perceived by the mechanosensors initiates a signaling cascade, which results in the growth of bone and inhibits local bone loss. The signaling pathway may include activation of protein kinases, regulation of G protein—coupled messengers or activation of secondary messengers such as Ca^{2+} and cAMP. A number of signaling pathways, which result in adaptive responses, exist in bone cells, of which a few are mentioned below.

1. Kinase signaling: Mitogen-activated protein kinase (MAPK) pathway is mechanically activated in bone cells. MAPKs are protein kinases, which are essential for cell proliferation and differentiation. ERK1/2 activation as a result of mechanical stimulus has been reported to be important in osteoblastic cells. ERK1/2 downregulates receptor activator of nuclear factor kappa B ligand and upregulates endothelial nitric oxide synthase in response to mechanical strain, which results in a lowering of the osteoclastic activity and hence enhanced bone modeling. ERK1/2 also brings about Ca^{2+} transport in voltage sensitive calcium channel via mechanical stimulus. Other kinases such as Akt can also be activated by mechanical stimuli and affect a number of cellular functions. Strain amplifies the mechanical activation of Akt or protein kinase B, resulting in the clustering of focal adhesions and may also regulate the development of cells in other tissues such as fat and muscle. Focal adhesion kinase (FAK) activation affects the formation of the FACs, which are responsible for further regulating integrin association and clustering that affect downstream MAPK pathways. FAK is essential in the upregulation of osteopontin and COX-2 expression in osteoblasts under oscillatory fluid shear stress, which increases osteocalcin, Runx2, and Osx expression, which are important indicators of osteoblast differentiation. Another focal adhesion signaling molecule, proline-rich tyrosine kinase (PYK2), has been shown to increase bone formation and reduce bone resorption. This same PYK2 also plays an active role in osteocyte apoptosis and hence may be important in mechanotransduction because this phenomenon has been observed to decrease under the conditions of loading.
2. Wnt signaling: Osteoblasts have shown the activation of the Wnt signaling pathway when subjected to mechanical strain. β-catenin is strongly upregulated in osteocytes in mechanically loaded bone. It is translocated to the nucleus in osteoblasts under the action of both fluid stress and mechanical strain. The critical role of low-density lipoprotein receptor—related protein 5 (LRP5) in many of the functions of bone tissues under mechanical stimulus was recently discovered.

Loss-of-function mutations in LPR5 lead to decreased bone density even if there is no appearance of abnormalities in the formation of bone. Studies show that the effectivity of exercise and the corresponding formation of bone mass is linked to the important role of LPR5 in mechanotransduction in the canonical Wnt signaling pathway. A decrease in sclerostin, which is an LRP5 antagonist, is associated with increased bone formation, especially in regions of bone tissue subjected to large strains, suggesting that sclerostin may be a strong modulator of LRP5 activity. It is unclear at present if LPR5 release from inhibition is sufficient or if augmented Wnt secretion is also conditionally necessary to transduce mechanical signaling through this pathway.

3. GTPases signaling: Mechanical stimulus has been reported to activate GTPases in the cell. This is mainly accomplished by the activation of G protein—coupled receptors. For example, protein kinase G is activated by NO, which is generated as a result of mechanical shear in osteoblasts. In addition, RhoA GTPases play a key role in the organization of actin stress fibers in response to mechanical strain. The fate of MSCs in terms of differentiation into either bone cells or adipocytes is influenced by RhoA based on modulation of cell shape. Another functional adaptation because of RhoA is cytoskeletal stiffening in response to a mechanical stimulus.
4. Calcium signaling: Calcium homeostasis is extremely important in all cells and especially in bone tissue formation. There is an instantaneous peak in intracellular Ca^{2+} concentration on the application of a mechanical stimulus, although such concentration is strictly regulated to be normally kept quite low. Thus Ca^{2+} serves as a key second messenger, and hence, a spike in its concentration is an excellent indicator of a transduction event such as differentiation, cell motility, etc. It has been reported that the frequency of Ca^{2+} concentration spikes is more revealing of the transduction process than the magnitude of calcium concentration. This lends credence to the greater importance of the oscillatory nature of the applied mechanical signal, compared with its magnitude. The variations in the concentration of intracellular Ca^{2+} are an indication of the activation of other signaling cascades as a result of mechanical stimuli, including MAPK, c-fos, and release of PGE2, which are all important in the formation of bone tissue. The spike in Ca^{2+} precedes the release of ATP, and any inhibition of Ca^{2+} results in decreased bone formation. The release of PGE2 and NO follow the above events, as evidenced by the presence of Cox-2 in immunostaining of bone tissue. The release of NO is another early indicator of mechanotransduction especially in osteocytes.

The signaling processes in bone cells have merited a more detailed description in this chapter because much more work has been done on this tissue. The results obtained from studying bone tissue are the basis for study of such mechanisms in other tissues. However, in the following sections, some emphasis will be placed on soft tissue, including cartilage, pulmonary, cardiovascular, and nerve tissues.

3.5.5 The response of bone cells to applied mechanical stimulus

A complex mechanical environment exists in bone tissue based on the mechanical stimuli experienced by the bone cells in a combination of stress, strain, pressure, fluid flow, streaming potentials, and acceleration. However, each of these factors is dynamic, and hence, constantly changing, although their individual magnitudes may be relatively small. It is also difficult to isolate and study to understand the individual effects of each mechanical stimulus because the response is combined and complex even for only one type of stimulus that is applied. Nonetheless, regardless of the type of applied stimulus, it appears that a continuous, high-frequency, low-magnitude mechanical stimulus is much more effective in producing strong functional adaptation in bone cells. For fluid flow, oscillatory, rather than laminar fluid flow, appears to be more effective in generating a more potent signal and seems to be the main reason for changes in cell characteristics. Besides, a large magnitude of a particular stimulus is not necessary to elicit a response from the cells, but rather short spurts of loading with adequate intervals of rest may be sufficient to trigger a response from the cells. On the contrary, rapid bone loss accompanies the unloading of skeletal tissue as observed from the drastic loss of bone tissue in astronauts under microgravity conditions.

3.6 Cell mechanotransduction in cartilage

3.6.1 Mechanical environment of cartilaginous cells

Cartilage, in general, and more specifically, articular cartilage, is avascular and aneural in nature. There is a paucity of chondrocytes in this tissue, and in accordance with the limitation in the transport because of avascularity, this tissue is not the subject of regeneration based on the lack of supply of nutrients. Hence, mechanical stimulus becomes a critical parameter and in once sense the hallmark for growth and development of this tissue. Cartilaginous tissue is highly hydrated with 75%–80% water. The rest of the tissue consists of 50%–75% of Type II collagen and 15%–30% of proteoglycans. Chondrocytes are the principal cell type in cartilaginous tissue. Immature chondrocytes are sometimes referred to as chondroblasts, although the precursors to these are MSCs, which can differentiate into chondroblasts or osteoblasts. Articular cartilage is subjected to extreme loading conditions both in compression and shear. Hydrostatic pressure also appears to have a strong influence on the physiological behavior of chondrocytes. Such mechanical stimuli have been implicated as key modulators in the process of regeneration of cartilage.

3.6.2 Mechanisms of mechanotransduction in chondrocytes

Fluid shear stress appears to have a positive effect on the synthesis of glucosaminoglycans in chondrocytes in culture, although hydrostatic pressure appears to have an even stronger influence. In general, similar to bone cells, intermittent or oscillatory mechanical stimuli at low frequencies (0.1−0.5 Hz) and small amplitude forces (<5 MPa) are reported to have the most significant effects on synthesis of aggrecan and Type II collagen. Three-dimensional constructs where chondrocytes are seeded are influenced much more by compression type loads. Even in these cases, intermittent compared with constant loading in compression is more effective in matrix synthesis.

The mechanisms proposed for mechanotransduction in chondrocytes involve the conversion of the mechanical signals into biochemical responses. Among these, the most popular are stretch ion channel signaling, integrin receptors, cytoskeletal elements, and deformation of the nucleus. Stretch-activated ion channels have been singled out as an important vehicle in the transport of Ca^{2+}, which is a universal second messenger in processes involving mechanotransduction. As mentioned elsewhere, intermittent transport of Ca^{2+} through stretch-activated ion channels in waves is in concordance with observations of oscillatory application of stimulus to produce the most effect in ECM synthesis in cartilage. Hydrostatic pressure up to a certain limit causes membrane depolarization in chondrocytes and affects the transport of Na^{+} and K^{+} in these cells.

Integrins, which are transmembranal proteins, act as mechanosensors in chondrocytes and can transmit the stimulus into the cell through the actin filament in the cytoskeleton. An integrin-regulated interleukin-4 (IL-4) pathway has been proposed for mechanotransduction in chondrocytes. This signaling pathway involves the activation of both stretch-activated ion channels as well as the chain of transduction events through the integrins. This is followed by the binding of the IL-4 signaling molecule to the receptor, which leads to a cascade of intracellular events causing the activation of potassium channel, resulting in hyperpolarization of the cell membrane.

The cytoskeleton also transduces mechanical forces in cells and brings about conformal changes, motility, and proliferation by reorganizing itself based on the nature of the applied stimulus. Mechanotransduction also causes changes in the actin, vimentin, and tubulin levels in cartilage based on the magnitude of the applied load. The deformation of the cell may also lead to nuclear deformation as a result of signal transduction and directly affect gene synthesis. The changes in nuclear structure have been reported to affect aggrecan biosynthesis.

3.7 Cellular mechanotransduction in pulmonary tissue

3.7.1 Types of cells in lung tissue

Lung fibroblasts, pleural mesothelial cells, and SMCs and epithelial cells in the airway make up the different types of cells in this system. The cells of the pulmonary artery wall, which are similar to the vasculature in the cardiovascular system can be considered as part of this system. The cells in the lung are subjected to a complex action

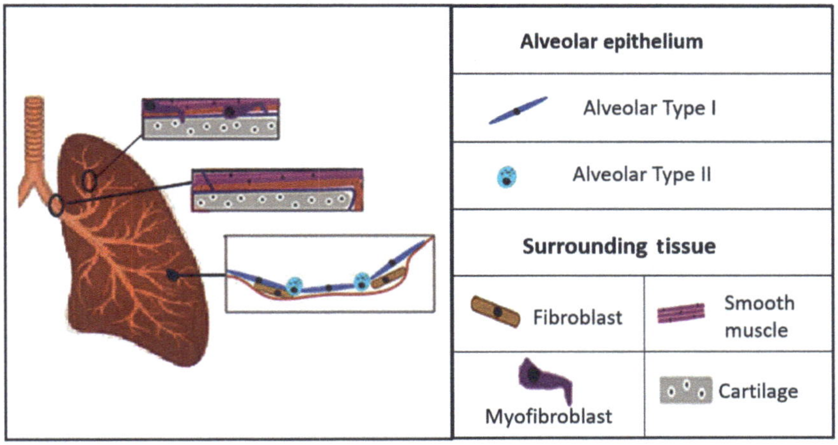

Figure 3.8 Cells in the pulmonary tissue. Types of cells in pulmonary tissue including fibroblasts, smooth muscle cells, cartilage cells, and epithelial cells (alveolar I and alveolar II cells). Other cells that are present in the pulmonary system are not shown.

composed of a variety of dynamic mechanical forces. Lung tissue includes the parenchyma, the air passageways leading up to the alveoli, and the vascular cells comprising of both pulmonary and bronchial blood vessels. Fig. 3.8 shows some of the cell types present in lung tissue. Although many other cells are present in this system only a few relevant cells are shown.

3.7.2 Mechanical environment in the lung

A variety of mechanical forces are pervasive in the functioning of this organ from the cyclic breathing pattern involving the inflation of the lung and the perfusion of material across the alveolar/vascular membranal interface. When abnormal or extreme physical activity is added to this scenario, the dynamic environment of the lung becomes even more complex. The constant distension and contraction are thought to be essential for the development of the lungs. In addition, the lungs are protected from collapse during expiration by the secretion of surfactant by the lung epithelium, which decreases surface tension. The cells in the lung experience a complex stress state because of the three-dimensional architecture in the lung.

All cells in the lung are strained to some extent because of the distention in this organ during breathing. The degree of stretch is probably greater in the lung epithelium and alveolar wall, whereas vascular endothelium is subjected to cyclic hydrostatic pressure and the corresponding strain. As in other systems, lung fibroblasts are subjected to a tensile mechanical stimulus, and the pleural mesothelial cells experience physical forces during the distension and inflation of the lung. Shear forces in varying magnitudes affect the vasculature from the larger pulmonary artery to the smaller capillaries and the pleural ECs as a result of fluid flow along the pleural wall and on the surfaces of the airway and alveolar epithelia. Shear forces act as a result of fluid movement between the parietal and visceral pleura during the mechanical operation of the lung while endothelin-1 (ET-1), which is a bronchoconstrictor is released under the action of this shear stress.

As a result of the mechanical operation of the lung, forces are generated within cells because of the conformal changes in the cytoskeleton. These forces can be transmitted to the nucleus through the microtubules and actin microfilaments. In short, the changes brought about by the forces transmitted by the cytoskeleton causes cell differentiation and proliferation. As in other specific cell types, the regulation of growth factors also occurs because of cytoskeletal alterations. External mechanical stimuli and forces generated by the cell are sufficient to bring about usual cellular processes into action, including apoptosis and remodeling of the ECM. In the case of the lung, these cell functions also involve surfactant metabolism. A few attempts have been made to quantify the mechanical forces experienced by lung cells, but the variation in the data is rather wide. Alterations in cell shape are induced by cyclic strain in airway SMCs, which assume an orientation perpendicular to the applied stretch and a more spindle-shaped configuration after mechanical stimulus. Airway epithelial cells exhibit upregulated ciliary beat frequency upon being deformed mechanically. As in the cardiovascular system, an increase in the arterial wall stress results in a thickening of the capillary basement membrane. Increased cellular wall stress results in overexpression of the mRNA content of various ECM proteins and growth factors. ECs in the pulmonary arterial wall are clearly subjected to shear stress, whereas the SMCs in the media and the fibroblasts in the adventitia are affected by deformation during distention and hydrostatic pressure. Pressure stimulus increases the deposition of matrix proteins in SMCs but not in the fibroblasts. Stretch rather than physical force appears to cause cell proliferation. The expression of endothelial factors occurs in a shorter time compared with the other cells, which take longer to synthesize and secrete matrix proteins. Cyclic stretching increases cellular proliferation in the pulmonary vasculature accompanied with an overexpression of βFGF, whereas constant pressure results more in increased protein synthesis.

3.7.3 Mechanisms of mechanotransduction in the lung

A number of mechanisms involved in mechanotransduction have been proposed for cells in the pulmonary system. The ECM—integrin—cytoskeleton pathway has been clearly elucidated for lung cells. The cells are connected to each other and to the ECM through a number of transmembrane cadherin and integrin receptors, respectively. These, in turn, are connected intracellularly via cytoskeletal plaque proteins, which form the FAC. FACs are groups of connector proteins, which form a macromolecular scaffold and basically couple the ECM proteins to the cytoskeleton. A strain-based mechanism on the transmission of forces through the cytoskeleton has been proposed, where actin filament—associated protein-110 transmits the force felt by the shape change in the cytoskeleton and activates c-Src tyrosine kinase. Another proposed mechanism is the activation of epithelial growth factor receptor (EGFR) in epithelial cells, which, in turn, activates ERK MAPKs resulting in the formation of surfactants. ERK1/2 regulated by the EGFR in response to the change in pressure in the wall causes the release of heparin-binding epidermal growth factor in the intercellular space, which, in this case, is the mechanosensor.

Thus, in a general sense, in the pulmonary cells, mechanosensing takes place at the focal adhesions through the integrin families and at cell—cell adhesions via the cadherin proteins. The cytoskeletal plaque proteins transform the mechanical stimulus into a biochemical signal through mechanotransduction via mechanoreceptors. For ECs in particular, another specific integrin initiates the signaling, which is then transmitted for the modulation of gene expression via the focal adhesions, c-Src, and MAPK pathways. Another proposed mechanism is the stretch-activated ion channel where the net influx of Ca^{2+} or K^+ ions results in cell depolarization. This process may occur as a direct result of applied deformation effecting the conformal change in ion channel activation or indirectly by G proteins, which are suggested to activate the ion channels. For shear stress—related stimuli, heteromeric and small GTP-binding proteins are implicated in the first or early stages of mechanotransduction. In fluid flow shear stress, PGI2, a potent vasodilator and antithrombotic agent, is released mediated by pertussis toxin—sensitive G protein. An increase in flow stimulates the release of NO and activation of K^+ channels, which are dependent on calcium/calmodulin upregulation. Thus phospholipase C appears to be involved in shear stimuli transduction in ECs.

The cell nucleus may also be perturbed directly by externally applied mechanical stimulus. The role of the nucleus in chromosome organization and DNA transcription is affected by the mechanical stimuli by opening up DNA molecule sites that are normally restricted and other regulatory proteins in the nuclear membrane or by altering nuclear pore size and the transport of mRNA from the nucleus to the cytoplasm. Consequently, gene expression and protein synthesis are directly affected.

Although a number of signaling cascades have been cataloged and various second messengers have been listed, three predominant pathways have been established. First, under shear stimulus and cyclic stretch, phospholipase C—mediated breakdown of PIP_2 and other phospholipases are activated. This results in the IP_3 and diacylglycerol upregulation, which, in turn, increases Ca^{2+} levels and activates protein kinase C (PKC). The higher Ca^{2+} levels activate exocytosis, whereas PKC may induce slower, long-term responses. Second, adenylyl cyclase is activated, which upregulates cAMP and protein kinase A activation, which has been implicated in compensatory lung growth. Third, the MAPK signaling pathway important for gene transcription is also activated through shear stress stimulus in ECs.

3.8 Cell mechanotransduction is the cardiovascular system

3.8.1 Types of cells in the cardiovascular system

The cardiac muscle includes fibroblasts, SMCs, and cardiomyocytes. Fibroblasts are responsible for the formation of the ECM; whereas the cardiomyocytes provide contractility to the atria and ventricles. Cardiac tissue is also made up of pacemaker cells, which generate electrical pulses and control the involuntary beating of the heart. The vasculature is lined with ECs and SMCs, which provide structural support to the vasculature. A variety of mechanisms control blood flow in the human body both under rest and under strenuous working conditions. Clearly, the functioning of the cardiovascular system is complex, and a coordinated effort is required from all types of cells present in this organ to carry out its primary physiological function of pumping and circulating blood to the different regions in the body. Fig. 3.9 indicates the different types of cells present in the cardiovascular system, which experience a complex mechanical stress environment per se.

3.8.2 Mechanical environment for cardiovascular cells

The physiology of the vasculature is strongly controlled by the fluid shear stress it experiences, which depends on the type (whether laminar) and rate of fluid flow. The

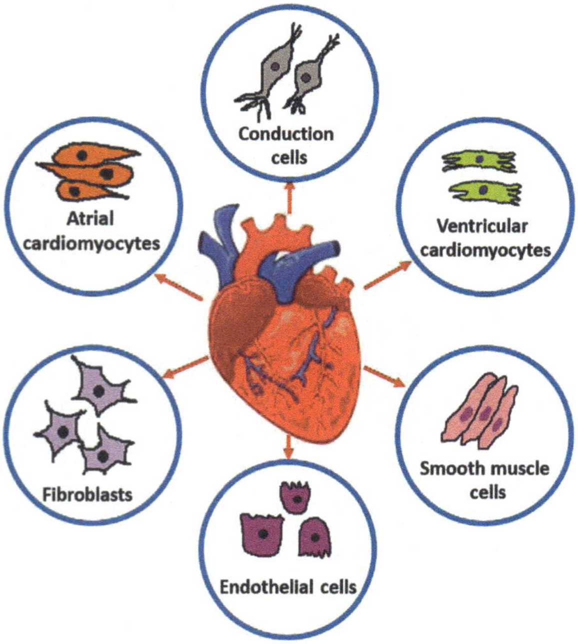

Figure 3.9 Cells in the cardiovascular system. Cells present in cardiovascular tissue maintain structural and mechanical integrity. A variety of cells that experience mechanical stimuli, including cardiomyocytes (both atrial and ventricular), smooth muscle cells, and endothelial cells, are shown.

Table 3.1 Transient responses of ECs to fluid shear stresses.

Duration	EC response
Seconds	• Activation of inward rectifying potassium channel • Release of prostacyclin • Phosphorylation of the PECAM-1 intracellular domain
Minutes to 10 minutes	• Activation of intracellular signaling molecules Ras, ERK, JNK, Rho GTPases, and tyrosine kinases
Minutes to hours	• Cell elongation • Actin cytoskeleton alignment • Orientation of microtubule organizing centers

EC, Endothelial cells; *ERK*, extracellular signal-regulated kinase; *JNK*, c-Jun N-terminal kinase; *PECAM*, platelet endothelial cell adhesion molecule.

myogenic response in the smooth muscle tissue in the arteries controls pressure appropriately to maintain the tissue in a healthy state both in the robust larger blood vessels as well as in the more delicate capillaries from pressure extremes by regulating localized flow. Based on the fluid flow, in the short term, SMCs in the arterial walls undergo relaxation because of high blood flow rates through the endothelial release of NO. For long durations, however, high flow rates result in blood vessel enlargement. Even implants such as vein grafts are thought to be arterialized through the expression of artery-specific genes based on flow patterns in the artery. Normally, unidirectional laminar flow is associated with the transport of blood in the arteries. However, disturbed flows that occur at vessel branching, curvatures, oscillatory flow, and other abnormalities in blood flow promote atherosclerosis in localized regions. Pathways, which are normally triggered transiently, are also activated in a sustained manner during disturbed flow. This is dangerous from a disease development perspective because cells can downregulate the signal during laminar flow to normal levels unlike in disturbed flow where the signal cannot be controlled. Prolonged exposure to elevated blood pressure leads to remodeling of the arteries and results in hypertrophy and cardiac failure.

Over various periods, a number of responses of ECs to fluid shear stress are observed, which operate in a matter of seconds to response that may take hours. Table 3.1 summarizes these responses succinctly.

Furthermore, the atrial and ventricular cardiomyocytes are constantly experiencing a pulsatile stress as the heartbeats.

3.8.3 Mechanisms of mechanotransduction in cardiovascular cells

Although it is universally acknowledged that cardiac muscle cells sense mechanical stimuli through the activation of the angiotensin II receptor, the mechanisms underlying this process are not well understood. Stretch-activated ion channels and integrin-mediated mechanosensors are responsible for the cellular response to mechanical forces in cardiomyocytes.

More specifically, melusin, which is a muscle-specific adhesion protein, is responsible for responding to mechanical stimuli. Rather than affect the tissue architecture, absence of this protein may indirectly influence mechanotransduction. Calcium transport and signaling ultimately play a critical role in this process through stretch-activated ion channels. Ion channels are activated through depolarization when the plasma membrane is stretched, allowing Ca^{2+} influx, which then regulates basal tone and myogenic constriction as a result of which Arg-Gly-Asp (RGD) peptides induce both a controlled vasoconstriction and vasodilation as part of this myogenic response. It is unknown if this vascular actuation is a result of integrin binding or decreased integrin adhesion to the ECM. Either way, integrins are highlighted as the mechanosensors when arteries are stretched, and when these proteins cluster for structural reinforcement, they activate FAK and MAPK. Overall, mechanical stretch upregulates integrin binding by increasing integrin affinity to the ECM.

In the case of ECs, the role of shear stress is extremely important. Shear stress modulates a number of endothelial functions, which control vasodilation, vasoconstriction, growth, adhesion, etc., by regulating the encoding of genes. For these cells, fluid shear stress is applied to their surfaces. Because a 300 nm thick glycocalyx covers these cells, there is no shear stress at the apical plasma membrane. Therefore, in this case, G proteins have been implicated in the transduction process, whereas both cell—ECM adhesions and cell—cell adhesion molecules are the mechanosensors. The applied force is transmitted to PECAM-1 via the cytoskeleton, which then activates PI3 kinase downstream. Cadherin molecules at the cell—cell junctions may be ultimately responsible for force distribution to the neighboring cells to perform cell physiological functions in a coordinated manner. Under the action of shear stresses, NOS is activated causing blood vessel relaxation. Prostacyclin, which is another powerful vasodilator, is also upregulated by the shear stress stimulus. In general, the effect of initial NO release relaxes the blood vessels. When this is combined with the decrease in monocyte adhesion to the EC layer, the result is antiatherogenic. Simultaneously, the levels of ET-1 are downregulated, decreasing vasoconstriction when ECs are stimulated with shear stress above a certain threshold value in the laminar flow regime. ICAM-1 expression is upregulated by shear stress encouraging leukocyte adhesion to ECs indicating that this mechanical stimulus may play a role in wound healing. COX-2, an antithrombotic enzyme and thrombomodulin, which activates anticoagulation are both upregulated by shear stress, highlighting its role in thrombosis and fibrinolysis. Oscillatory or sudden changes in shear stress appear to increase proliferation by expression of *c-fos* and *c-jun* genes. However, this is absent under sustained shear stress application. Pulsed shear flow is more effective in affecting various growth factors involved in the proliferation of ECs. The magnitude of laminar flow shear stress correlates with the expression of genes above a threshold of 5 $dynes/cm^2$. However, there also appears to be an upper limit for this correlation.

There are a number of shear stress—induced responses of signaling molecules involved in mechanotransduction in ECs. These include MAPKs, small GTPases, heterotrimer

G proteins, and protein tyrosine kinases (PTKs). Application of shear stress rapidly activates ERK and c-Jun N-terminal kinase (JNK) in ECs, and this transient effect results in *c-fos* expression by ERK and *c-jun* phosphorylation by JNK. The equilibrium in the activity of these two genes is essential for maintaining adequate cell growth and balanced apoptosis. The small GTPase molecule Ras is vital for cell proliferation, and similar to the other MAPK molecules mentioned above are rapidly activated under the action of a shear stress on ECs. The Rho family GTPases, on the other hand, are linked to the formation of focal adhesions. Shear stress also regulates the formation of stress fibers and cell shape by activating PTKs and the early stages of NO production. Of the two categories of PTKs, the nonreceptor tyrosine kinases (RTKs) FAK and c-Src are confined to only cytoplasmic activities because they have no extracellular domains. It should be pointed out that enhanced tyrosine phosphorylation of both FAK and c-Src occurs in the presence of shear stress.

Temporal and spatial responses to gene expression in ECs are very important for maintaining healthy vascular tissue. The variation in times necessary for activation of the different signaling molecules in the ECs points to a time sequence in their signaling based on their function in the cell. Membrane molecules such as Ras and c-Src respond to activation in less than a minute and reach peak activity in about 5 minutes. Nonreceptor PTKs are activated more slowly in a time frame of 10–30 minutes, whereas transcription factors translocate to the nucleus over a period depending on their role. Intermediate early genes are activated in minutes, whereas MAPK genes such as *c-fos* and *c-jun* take about 30 minutes to initiate activity before reaching a peak in 1–2 hours. Fig. 3.10 describes the periods for activation and peaking for signal transduction and gene expression in ECs under the action of a shear stress.

Comparative plots of the temporal EC response under the action of a shear stress to signal transduction and gene expression, respectively. The bars indicate the duration

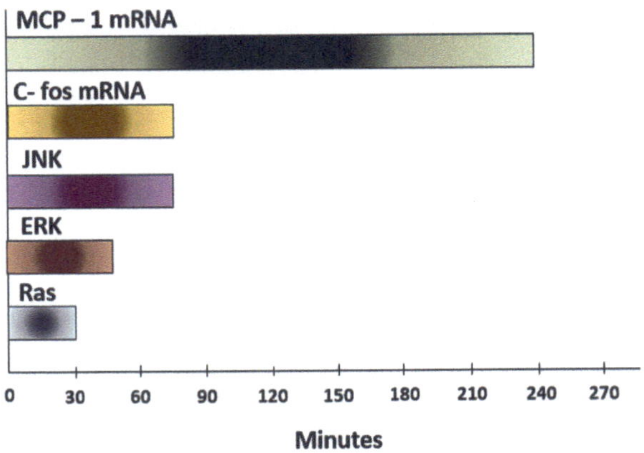

Figure 3.10 Temporal endothelial cell response. *ERK*, Extracellular signal-regulated kinase; *JNK*, c-Jun N-terminal kinase; *MCP*, monocyte chemoattractant protein.

of the signal transductions. Ras is activated instantly, while ERK and JNK show delayed reaction. Corresponding reactions of encoding genes c-fos and monocyte chemoattractant protein-1, which exhibit activation peaks later are also depicted. The intensity of each color reflects the level of activation for each case with peak activation corresponding to the highest intensity of color. Basal activity (not depicted here) in this case corresponds to the static control threshold for similar activity.

The ECs in the vasculature are also subjected to circumferential stress as a result of blood flow. Therefore these cells are constantly experiencing a cyclic strain or stretch. Again, intermittent rather than a sustained load signal is more conducive to normal gene expression.

3.9 Cell mechanotransduction in nerve tissue

3.9.1 Types of cells in the nervous system

Generally, the cells in nervous tissue can be categorized into two broad groups: neurons, which are the information processors, and glia, which regulate the metabolic processes in the neurons and also provide mechanical support. Neurons are classified into (1) receptors whose function is to encode sensory information to transform sensory touch to electrical and chemical signals, which are interpreted and transmitted to other cells; (2) interneurons, which simply relay the signals which they receive; and (3) effector or motor neurons, which regulate muscular or glandular activity by sending signals to these tissues. Of these, interneurons make up the bulk of the nervous system.

A single neuron can be divided into the cell body, dendrites, and axon. The nucleus is housed by the cell body, whereas the dendrites are basically cell body extensions, which receive information and transmit them to the cell body. Many neurons have axons, which is the principal extension that relays information from the cell body to other cells. Neurons are connected to each other and thus transmit information over relatively long distances. The region of communication between one neuron and the next is called the synapse. The signal travels from the endfeet, which is the terminating extensions of the axon of a transmitting neuron called the presynaptic cell, to the axon of the receiving neuron called the postsynaptic cell. Dendrites and axons are extensions of the cell body and are generally termed processes. Based on the number of processes, neurons can be unipolar, bipolar, or multipolar, the latter being typical in vertebrate systems. Neurons with long axons are called principal axons and transmit information over relatively large distances within the brain. On the contrary, localized circuit neurons only influence their immediate vicinity. Correspondingly, principal neurons have rather large cell bodies compared with the localized circuit neurons.

The second group of cells, which are important in the neural tissue, are the glia, which are interspersed with neurons, although they consist of more than half the weight of all the cells in the nerve tissue. Glia are smaller cells and are 10–50 times more in number than neurons. Two types of glial cells, the larger macroglia, and the smaller

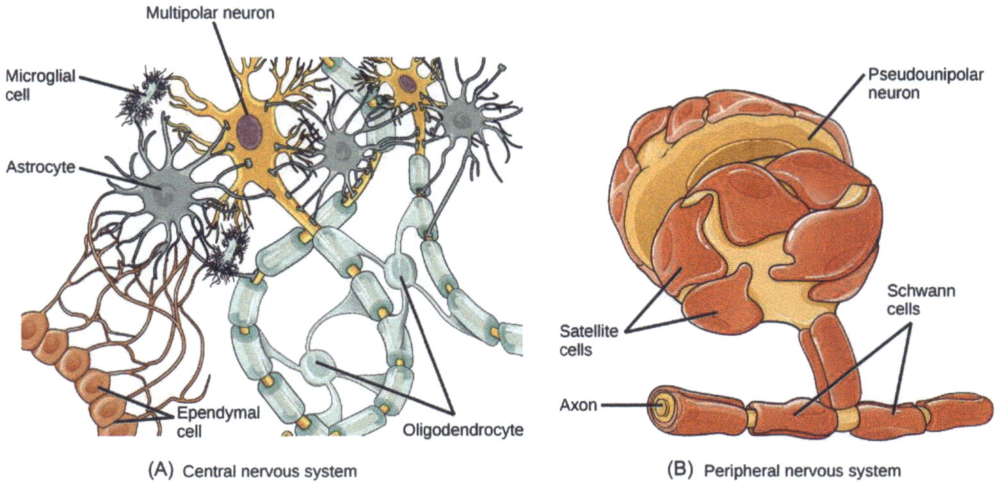

Figure 3.11 Types of nerve cells. Cells in both central nervous system (CNS) and peripheral nervous system (PNS) including neurons and glia. Multipolar neurons and microglial astrocytes and oligodendrocytes are shown in the CNS. Microglia, which clean up the nerve tissue, are also illustrated. Schwann cells that myelinate the axons are shown in the PNS (CC BY 4.0).

microglia are present in nerve tissue. Among the macroglia are two types of cells called astrocytes, so named because of their star shape in stained tissue and oligodendrocytes. Astrocytes principally provide structural support to neurons and possibly aid in the repair of damaged neurons. On the other hand, the primary task of the oligodendrocytes is to produce myelin, which forms the protective insulative coating called the myelin sheath, which surrounds the axons. Another important cell type in the nerve tissue are the Schwann cells, which play a crucial role in the myelination of the axons. These insulate the axons, which is extremely helpful in rapidly transmitting the action potentials in the axon. Microglia, which account for about 10%–15% of all the cells in the brain, play a more secondary role in the nervous system by eliminating dead cells from the tissue. Fig. 3.11 shows the different types of cells present in the nervous system.

3.9.2 Mechanical environment for nerve cells

Under normal physiological condition, the peripheral nervous system constantly experiences tensile, compressive, and shear stresses as a result of locomotion and limb movement. Generally, mechanical stresses are also imposed on nerve tissue during development. The proximity to other tissues such as muscle and tendon also results in the nerve cells being compressed by the extraneural environment. Neurons and Schwann cells are very sensitive to the stiffness of the surrounding nerve microenvironment. Other than integrating the signal received by the axon and transmitting it, the cell body also has metabolic tasks in providing the neuron with the energy to carry out its physiological functions. It performs a number of biochemical processes necessary to synthesize proteins required for signal transmission. Very importantly, the neurotransmitters that are released in the synapse are produced in the cell body and transported along the length of the axon. Depending on the length of these axons, they are stretched during growth and maturity and hence experience axial tensile and transverse compressive stresses. The axon sprouts out in the form of a conical membrane from the cell body called the axon hillock and consists mainly of microtubules and microfilaments whose primary goal is to transport substances through this system before releasing them in the synaptic cleft. As such, the microtubules and microfilaments have influential mechanical properties, which clearly affect the neural microenvironment.

3.9.3 Mechanisms of mechanotransduction in the nerve cells

The understanding of the influence of mechanotransduction in nerve cells is still at a nascent stage. Integrin adhesion complexes (IACs) are the fundamental mechanosensors and mechanotransductors in neuronal cells. These adhesive structures are highly complex and sophisticated enough to sense the mechanical cues in the cell microenvironment and transmit these signals further into the cell. These are the principal mechanosensors although others such as glycosylphosphatidylinositol-anchored proteins and RTKs collaborate in this process. Eighteen heterodimeric structures make up the integrin family with 18 α and 7 β subunits, which have resulted in 24 combination units up to now. Most of these have large extracellular domains, which provide various binding domains to the ECM proteins offering a wide palette of sensors, which lead to a number of distinct functionalities of the integrin-mediated mechanotransduction. The activation of these integrin sensors is accomplished through a number of biochemical means but also through force signals external to the cell, which affects the cell membrane. Such activation exposes binding sites in the cytoplasmic tail of the β subunit where talin and kindlin attach. Another mechanism is the compression of the glycocalyx close to the integrin—ECM interface, which results in the transmission of force through the integrin and the cell membrane. Through a mechanism, which is still not adequately elucidated, the talin binds to actin filaments. Based on the magnitude of the force acting in this structure, cryptic binding sites are unfolded in talin to which the vinculin head attaches itself, whereas the vinculin tail forms a catch bond with the actin filaments. As a result of this force transmission up to the cell nucleus, nascent focal adhesions are strengthened and ultimately leads to the formation of integrin clusters, which are critical in cell migration and proliferation. The size of the IACs depends on their role in the mechanophysiological functions of the cell, including cell signaling and cell fate. Thus the interactions at the cell/microenvironment interface are extremely important for regulating appropriate cell behavior, the lack of which leads to many pathological conditions and disease.

The composition and topography of the ECM in the neural tissue and the sensing and transductional processes as a result of mechanical stimulation at the integrin interface are critical in the control of neuronal cell development and function. The interactions of integrin with its microenvironment along with Rho GTPase signaling clearly affect the migration of neuronal progenitor cells as they move to their terminal destinations to differentiate into neurons. Neuronal progenitor proliferation in the stem cell niche is dependent on MAPK signaling, which is activated by integrin-mediated laminin binding. The mechanosensory yes-associated protein encourages neuronal progenitor proliferation but deters their differentiation. Neuronal migration requires tight control of Rho GTPase signaling, actomyosin activity, and traction forces. The loss of Rac1 leads not only to apoptosis of neural progenitors and incipient neurons but also produces defects in migratory ability, axonal guidance, and terminal differentiation of neurons. A lower RhoA activity, on the other hand, positively affects the commitment of progenitors to neuronal lineage. Growth of retinal ganglion cell axons is directed toward softer regions of the brain tissue, whereas directional collective migration of neural crest cells can be controlled by varying the stiffness of the underlying tissue.

As one can clearly notice, the mechanisms of mechanotransduction operating in the above cells are similar although the final physiological response is specific to each cell type. As our understanding of this phenomenon increases, we may be able to distinguish nuances between these mechanisms, which are specific to each cell type.

Review and Project

1. What is the purpose of a mechanosensor in a cell? Enumerate the various mechanosensors and relate the cell function accomplished by each mechanosensor in various tissues.
2. Write down five signaling mechanisms used in mechanotransduction. Draw simple diagrams to track the possible signaling cascades for each one. Project on the deficiencies that can be produced if some of the phases in the signaling cascade do not operate.

CHAPTER 4

Cell characteristics resulting from mechanical stimulus

4.1 Introduction

Many studies have been performed in understanding the effect of mechanical stimulus on the characteristics of cells. Based on many such studies, the role of many proteins and other molecules at both the cellular and molecular levels in regulating processes in mechanotransduction has been clearly recognized. Although the knowledge gleaned from this research has been far-reaching, the depth of understanding of the corresponding phenomena is quite limited. If the vast number of parameters are factored into the effect of mechanical stimulus on cells cultured in vitro, the interpretation and analysis of the results obtained are complex. If the influence in vivo is added, the scenario becomes even more intractable. This chapter is divided into comprehensive sections beginning with the effect of mechanical stimulus on normal cells with an emphasis on resulting cell physiology and cell behavior. This will be followed first by a section on the influence of mechanical stimuli on stem cells and then on cancer cells. It is then important to cover the role of cellular adhesion structures on cellular sensing, the proliferation of cells under the action of mechanically applied loads or deformation, and finally, what, if any effects does mechanical stimulus have on the remodeling of the extracellular matrix (ECM).

4.2 Effect of mechanical stimuli on normal cells

4.2.1 Types of mechanical stimuli

For studies on the effect of mechanical forces or stretch on cells in 2D, usually in vitro systems involving flexible substrates are employed to culture cells to carry out experimentation. Although the objective is to mimic the stimuli in the in vivo environment, realistically, it is impossible to achieve such conditions. Flexible elastic substrates coated with substances that represent the ECM proteins, such as fibronectin, laminin, or collagen, have been effective at least in the application of stress albeit indirectly to fibroblasts, smooth muscle cells (SMCs), and endothelial cells (ECs) despite the fact that the loads applied to the substrate are not the same as those experienced by the cells. For chondrocytes, the application of the stimulus is slightly more complicated, as these are subjected to both a compressive stress and a hydrostatic stress component under in vivo conditions, and hence, the experimental system

Mechanotransduction
DOI: https://doi.org/10.1016/B978-0-12-817882-9.00003-3

for load application is a little more complex. For vasoendothelial cells, a fluid shear stress is more apt, and normally microfluidic flow chambers are utilized.

Uniaxial stretching is appropriate for aligned cells from tissues with a predominantly longitudinal geometry, such as tendons and ligaments. Dermal fibroblasts, on the other hand, experience random stretching for which controlled biaxial testing may be more useful. Osteocytes may be subjected to a purely compressive load, whereas vasoendothelial cells experience an apical shear stress, which is then translated to the basal laminin. Some precautions to keep in mind during such experimentation is that not all cells in the culture may experience the same type or magnitude of stimulus, especially when they are seeded randomly and are not adhered in the same manner on the substrate. As mentioned earlier, such heterogeneity sometime leads to ambiguity and variations in the interpretation of the resulting data. Cells oriented in a certain specific direction are preferred to add more uniformity to the results. This is accomplished by self-assembly of cells in microgrooves or surfaces with oriented topography.

A number of factors play crucial roles in affecting cell characteristics as a result of an applied mechanical stimulus. The following factors and the associated effects provide some insight into the influence of mechanical stimuli on cells. There is a wide-ranging span in the studies of mechanical stimuli on various cell types and the manner in which each cell line responds during mechanotransduction. In spite of the efforts to homogenize cell culture conditions and strictly adhere to the parameters applied during the laboratory experiments, the results obtained have shown the wide variety in the observed phenomenon and hence the difficulty of arriving at sound scientific conclusions. Nevertheless, it is a well-established fact that mechanical load has an effect on the growth and modeling and functioning of different tissues. Clear examples including the remodeling of bone tissue through mechanotransduction in osteocytes and the well-defined orientation of vascular ECs and structure of the lumen because of oscillatory loads have been documented. On an intracellular level, gene expression resulting from direct load application on the nucleus and cell reorientation and motility from applied mechanical stimulus have confirmed the effects of force application on cells in vitro.

4.2.2 Influence of mechanical factors

In a general sense, mechanical factors strongly affect the physiological behavior of all cell types.

Many cells increase their rate of growth and show increased proliferation under externally applied mechanical stimulus. Bone cells, on being mechanically stimulated at low strains but high frequency, deposit an increased amount of ECM. Compressive loads influence chondrocytes by altering the production of proteoglycans. The two most common modes of force application on cells in vitro are force applied by stretching of substrates on which cells are cultured or shear forces on cells because of fluid flow. The former usually affect cell adhesion proteins such as integrin connected to the substrate via ECM proteins or cadherins, which facilitate cell—cell connections. Fluid flow generates shear forces that can deform or stretch the cell membrane or act through membrane surface receptors such as primary cilia to accomplish mechanotransduction. Fluid shear stress is also very effective in influencing ECs to express proteins involved in binding and tissue remodeling through the activation of certain gene promoters.

As mentioned in the Chapter 3, there is a notion in the scientific community that under normal conditions, cells generate a contractile residual force, as they adhere to the substrate, and hence, there exists a tensegrity structure constructed by the cytoskeletal elements in the cell. This residual force equilibrium can be overcome by the application of a certain threshold force, which would then bring into play various physiological functions of the cell such as mitosis and cell migration. Only minor changes are sufficient to perturb the equilibrium of this tensegrity and produce the expected physiological changes, both in vitro and in vivo without the help of biochemical factors such as hormones. The contractile residual stresses generated by cells and their interaction with the substrate also regarded as critical factors in the adhesion process. If the substrate is soft, the cells are unable to obtain sufficient physical resistance to attach to it, become stable, and spread out. A stiffer substrate, on the other hand, is logically the preferred surface for attachment. Our understanding is still poor regarding the magnitude of these stresses or strains in the types of cell lines and their preferences. Neither is the range of forces or thresholds that trigger cell motion clearly delineated. Such knowledge would be critical in understanding cancer cell metastasis, and if indeed, such cells would migrate from softer breast tissue to preferential attachment and growth on stiffer substrates such as lung or bone and what the driving force to accomplish this type of cell migration.

4.2.3 Effect of substrate characteristics

Compliance or stiffness of the substrate exerts a sizable influence on the adhered cells even in the absence of regulatory factors. Substrate or ECM stiffness has a strong effect on cell physiology and function, and ECM mechanical properties may have an effect on cell fate. It is now accepted that in general, cells prefer to adhere to stiffer substrates or structurally modified ECM instead of the softer matrices. At the molecular scale, stiffer substrates or an ECM modified by other structural proteins to increase its elastic modulus cause the agglomeration of heterodimeric integrin proteins upregulated by an increasing mechanical force applied on the cell. In a similar manner, cells show a bias in motility directed toward stiffer substrates. One perspective is that this effect of stiffness favoring adhesion or a preference for migration toward a stiffer substrate may be localized. There is speculation that the migration of tumor cells to other tissues in the case of metastasis may be a result of an attempt by such cells to adhere to a stiffer substrate. This concept of matrix stiffness controlling cell attachment and motility has been termed *durotaxis*.

Other cellular physiological functions are also attributed to mechanical forces generated by a stiffer substrate. Cell growth, mitosis, differentiation, and even apoptosis can be regulated with minute alterations in the substrate compliance. A decrease in the stiffness of the substrate or ECM can cause many types of epithelium to change from growth mode to differentiation. The remodeling of the ECM and changes in its structure because of increased matrix metalloproteinase (MMP) activity is sufficient to produce changes in tissue growth and development. The shape of vascular SMCs can be changed by aptly designing the compliance of the ECM, which then enables them to regulate their contractile response to vasoagonists. Focal adhesions appear to grow faster on stiffer substrates and elongate in the direction of the applied stress in the actomyosin machinery. Specific signaling cascades are activated for cell division, migration, and apoptosis by the agglomeration of focal adhesions interacting with ECM molecules through transmembranal integrin. The aggregation of cell—ECM contacts may be a reaction of the cell to the application of an increasing applied mechanical stimulus, a clear example of mechanotransduction. Cell growth during the G1 to the S phases is also adjudicated to mechanotransduction on the application of a physical force on the cell. This upregulation of cell growth is reported to occur faster on stiffer substrates along with the alignment of cells in the direction of the applied mechanical load. It must be recognized that the response to an applied stimulus is not always instantaneous but can be prolonged over a duration of hours such as gene expression.

It is difficult to predict the minimum or threshold force magnitude that will elicit a cellular response. Bone formation, in comparison to cartilaginous tissue production, depends on the magnitude of the applied mechanical force, although exact values for this transition has not been defined yet. However, it appears that a threshold autobaric stress must be overcome for the cell to react to the applied stimulus. The magnitude of this stress varies from large values for cells comprising the musculoskeletal system in comparison to vasoendothelial cells, which respond to very small values of shear stress. The type of load experienced also varies from uniaxial modes of deformation in the longer limbs to complex hoop stresses in vasculature.

4.2.4 Effect of scaffold properties

For in vitro cultures, two-dimensional and three-dimensional scaffolds elicit differing responses in cell physiology and function. Clearly, cells cultured in 3-D do not exhibit apical polarity. However, these scaffolds appear to conform the seeded cells to the alignment of the underlying matrix fibrils. This effect is exacerbated under cyclic stretch much more than under static stretch. Whether this effect is a result of the applied mechanical stimulus or a natural reaction of the cells to adjust to the substrate is still moot. In 3D collagen gels, capillary outgrowth is observed when tensile stretch is applied. Similarly, nerve cells extend their processes in the direction of applied tensile stress. Altering the tension in the cytoskeleton can influence the morphogenesis of lung branching, as well as angiogenesis in both wound healing and tumor progression. ECs can be programmed to switch from growth to apoptosis. Placing restrictions on the space in which these cells are allowed to grow causes cell contraction or rounding and leads to apoptosis.

4.2.5 Mechanotransduction at the molecular level

Based on the discussion in earlier chapters, mechanotransduction can be considered as the connection between the mechanosensory stage and the eventual response of the cell through gene expression and protein synthesis. Various signal transduction pathways are activated as a result of the mechanical stimulus. Although the mechanism of mechanotransduction is not yet clearly elucidated, under cellular stretch mechanosensory responsibilities are relegated to integrins and cadherins. Primary cilia play a critical role in sensing mechanical stimuli under fluid flow producing shear, for example, in the role of remodeling during repair in bone tissue, whereas the endothelium is responsible for increasing the synthesis of ECM for repair or reinforcement of vasculature. Signal transmission becomes the main task of the actomyosin machinery, which acquires the signal from the complex group of adhesion proteins, including vinculin, talin, paxillin, zyxin, p130Cas, Src, and focal adhesion kinase (FAK) among others. It is generally accepted that mechanotransduction through integrin cell receptors is contingent to the formation of focal adhesions, which is then relayed to the nucleus through the cytoskeleton, especially the actin microfilaments. At the molecular level, some studies have suggested that vinculin is activated by the mechanical signal and acts in conjunction with proteins in the focal adhesion complex (FAC) to relay the stress signal. Talin may even be an upstream activator and possibly recruit vinculin to act bilaterally in the formation of stress fibers, in turn regulating the *Rac* and *Rho* GTPases. However, it is clear that the focal adhesions are mechanosensory and appear to self-assemble, grow, and align on the application of a mechanical stimulus on the cell. A reasonable deduction based on these observations is the fact that focal adhesions must disassociate and reform during cell migration. The conditions under which these complex proteins form and disassemble are not well understood. It is unknown at the present time if there is an equilibrium count in the density of focal adhesions per cell and their average size or if these vary primarily with substrate stiffness, topography, cell type, and intensity of the applied mechanical stimulus. In contrast, it is accepted that focal adhesions disassemble on softer substrates while they grow and mature on stiffer substrates aiding cell motility. Activation of RhoA because of applied mechanical stimulus is attributed to growth and maturation of focal adhesions. An interesting conjecture is that this process, in turn, could initiate and maintain stress-activated actin polymerization. Although it can be speculated that the number and size of focal adhesions increase with a larger magnitude load stimulus, there is no correlation to this fact and, at the present time, can only be conjectural. However, experimental evidence points to the maturation of nascent focal adhesions once a constant external mechanical stimulus is applied. Fig. 4.1 provides a clear

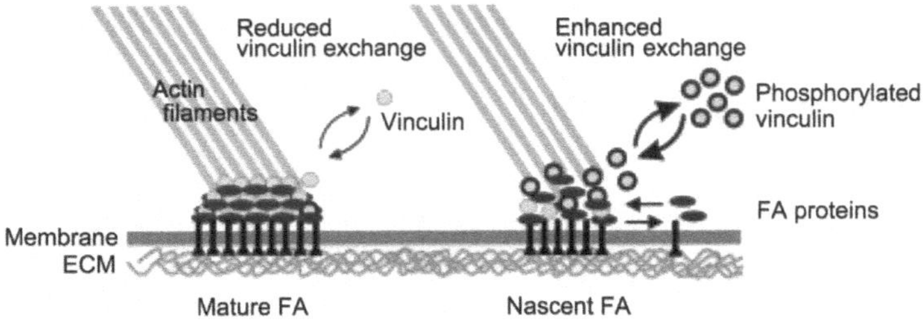

Figure 4.1 Growth of focal adhesions. Schematic illustration of the growth of nascent focal adhesion to maturity. Vinculin plays a critical role in acting as the adaptor protein to enable the transmission of the mechanical signal sensed by integrin to the actin stress fibers through the focal adhesion complex. This activates clustering of the focal adhesions and provides the cell with more stability. *ECM*, Extracellular matrix; *FA*, focal adhesion. *Möhl C., Kirchgeßner N., Schäfer C., Küpper K., Born S., Diez G., Goldmann W.H., Merkel R., Bernd H., 2009. Becoming stable and strong: the interplay between vinculin exchange dynamics and adhesion strength during adhesion site maturation. Cell Motil. Cytoskeleton 66, 350–364.*

illustration of the growth and maturation of focal adhesions based on vinculin exchange dynamics which appear to be controlled by applied mechanical stimulus.

The result of an applied mechanical stimulus is the activation of various signaling pathways, some mechanical but many biochemical. These pathways were briefly mentioned in a general sense in Chapter 2, Cell Signaling: An Overview, and described in much more detail in Chapter 3, Cell Signaling Under Mechanical Stimulus in Specific Cell Types. Some additional details of these pathways are provided here. The specific number of signaling pathways for mechanotransduction is currently unknown, although it is believed that many of these are shared despite differing end results in cell response or function. Normally, activation of these biochemical routes is through the usual second messengers and resulting in gene expression and protein synthesis. But, as mentioned earlier, direct mechanical effects are possible by stretch-activated ion channels, deformation of the cell membrane, and consequently, conformal changes produced in the proteins located in the membrane such as G protein—coupled receptors to direct translocation of proteins into the cell nucleus as a result of force projection through the actin microfilaments. Independent of the cell line, three principal mechanisms of mechanotransduction have been identified. These are mentioned below.

4.2.5.1 Integrin-based mechanism

Integrins are the first proteins in the cell to sense mechanical stimulation. In adherent cells, these transmembranal proteins are attached physically to the fibrous proteins, such as collagen and fibronectin in the ECM as mechanoreceptors on the one hand while being connected to the cytoskeleton on the other side. Application of mechanical forces activates mechanosensing by the integrins and promotes their clustering, which subsequently initiates mitogen-activated protein kinase (MAPK) signaling. The sensing mechanism may be because of the FAC undergoing conformal changes by positional exchange of the component proteins or by a shape change of these FA molecules. For example, vinculin and other proteins exist in active (open) or inactive (globular) forms and hence respond to the external physical stimulus by becoming activated and triggering the subsequent signaling cascade. Transmission of the signal continues through the actin stress fibers. Regulation of this tensile force occurs through phosphorylation of the myosin light chain, and the growth of stress fibers is regulated by RhoA, a small GTPase molecule, through its effector protein ROCK, which stabilizes F-actin and hence the stress fibers. In the absence of mechanical stimuli, remodeling of the cytoskeleton would be inhibited, and as a result, there is downregulation of gene expression. The influence of integrin-mediated signaling is deemed to be more localized. Under the action of a mechanical stimulus, the integrins normally begin to cluster to counteract the imposed external perturbation as an initial response. A second action is to make available new binding sites for other proteins resulting from conformal changes in response to the external signal. Both configurations result in the lowering of the free energy and hence are more stable. Fig. 4.2 illustrates the transduction mechanism for integrin-mediated mechanosensors.

4.2.5.2 Stretch-activated ion channels

A second mechanism suggested is the activation of ion channels by the external mechanical stimulus to upregulate the influx of Ca^{2+} ions. The stretch-induced influx of calcium ions leads to the activation of molecules such as nuclear factor kappa B (NF-κB), CREB, and other kinases. These, in turn, regulate downstream MAPK signaling pathways and increase COX-2 through nuclear translocation of NF-κB, which is a transcription factor and plays a direct role in many cell functions. Stretch-activated ion channels are considered to play a more global role in cellular signaling. The two possible mechanisms for the operation of stretch-activated ion channels based on free energy principles are described graphically in Fig. 4.3. The ion channels occupy a greater area in the bilipid membrane from mechanical stretching, which is a low free energy configuration and is preferred because it results in more stability.

4.2.5.3 G protein—coupled receptor

This mechanism is activated in ECs under the influence of fluid flow shear stress stimulus. Recent observations shed light on the conformal changes produced in fibronectin where certain cryptic sites are exposed under the action of fluid shear stress, which is transduced to a biochemical response by altering relative the distance between adjacent receptor binding sites. Such conformal change in the proteins or protein unfolding also results in a negative change in free energy as a result of which this response under the action of an external mechanical stimulus is favored. This is indicated in the illustration in Fig. 4.4.

Although in medicine and biology, biochemical factors are known to be the primary drivers in gene expression and protein synthesis, mechanical factors have been

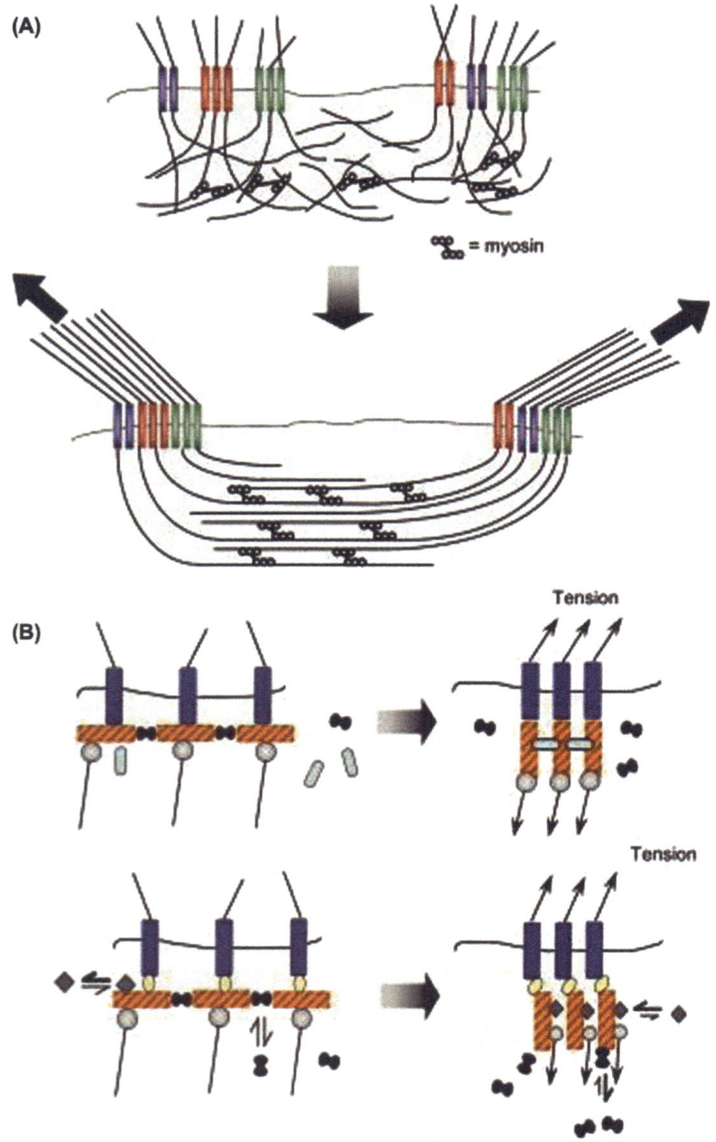

Figure 4.2 Operation of mechanosensorial integrin protein. The application of a mechanical stimulus results in (A) the clustering of integrins or (B) the exposure of new binding sites for other proteins. Both are the results of mechanotransduction with integrin acting as the mechanosensor/receptor. *Orr A.W., Helmke B.P., Blackman B.R., Schwartz M.A., 2006. Mechanisms of mechanotransduction. Developmental Cell. 2006; 10, 11–20.*

recently recognized as also playing a dominant role in these processes in various cell types, which are normally submitted to larger scale mechanical stimuli in vivo in connective tissue, including skin, muscle, and bone. Cells respond in a variety of ways to

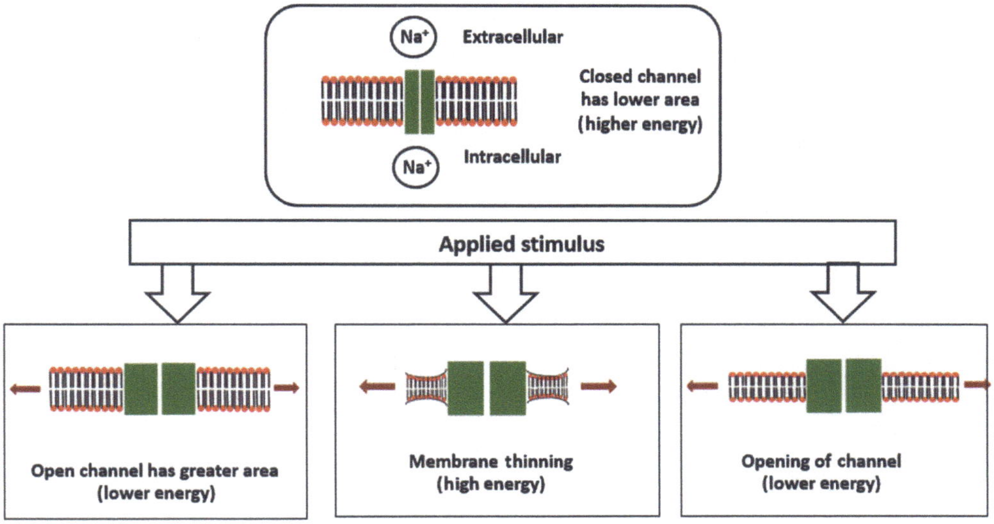

Figure 4.3 Operation of stretch-activated ion channels. Free energy-based operation of stretch-activated ion channels where the free energy of (A) the open state is lower than that of the closed state and (B) the open state is preferred over the closed state to avoid hydrophobic mismatch corresponding to increased energy. *Orr A.W., Helmke B.P., Blackman B.R., Schwartz M.A., 2006. Mechanisms of mechanotransduction. Developmental Cell. 2006; 10, 11–20.*

the applied mechanical stimulus. There is some controversy regarding the preference of cell orientation under the effect of applied stretch. Cells appear to orient themselves in a direction parallel to that of the applied mechanical load or strain in some cases, whereas in others, cells align themselves perpendicular to the applied stretch direction. Cell alignment has been forced through *haptotaxis* by engineering the topography of the substrate through the fabrication of microchannels on polydimethylsiloxane (PDMS) surfaces in in vitro studies. Thus cells are able to sense their microenvironment and orient themselves during adhesion to substrates by contact guidance and align themselves to microgrooved patterns. Mechanical forces are also capable of aligning the cytoskeleton in the direction of the applied force and also distort the nucleus and redistribute the nucleoli.

Collagen is structurally the strongest protein in these connective tissues and the most abundant protein in the human body. Collagen is expressed in greater abundance in the ECM under the action of a mechanical stimulus on the cells producing it. Both, collagen Type I mRNA and collagen Type III mRNA levels are upregulated under the action of stretching forces, whereas in wound healing and tissue regeneration in vivo, early onset of collagen Type III synthesis is followed by a more sustained collagen Type I deposition. Type I collagen is also expressed by fibroblasts subjected to strain on laminin or elastin. Glycosaminoglycans (GAGs) and proteoglycans (PTGs)

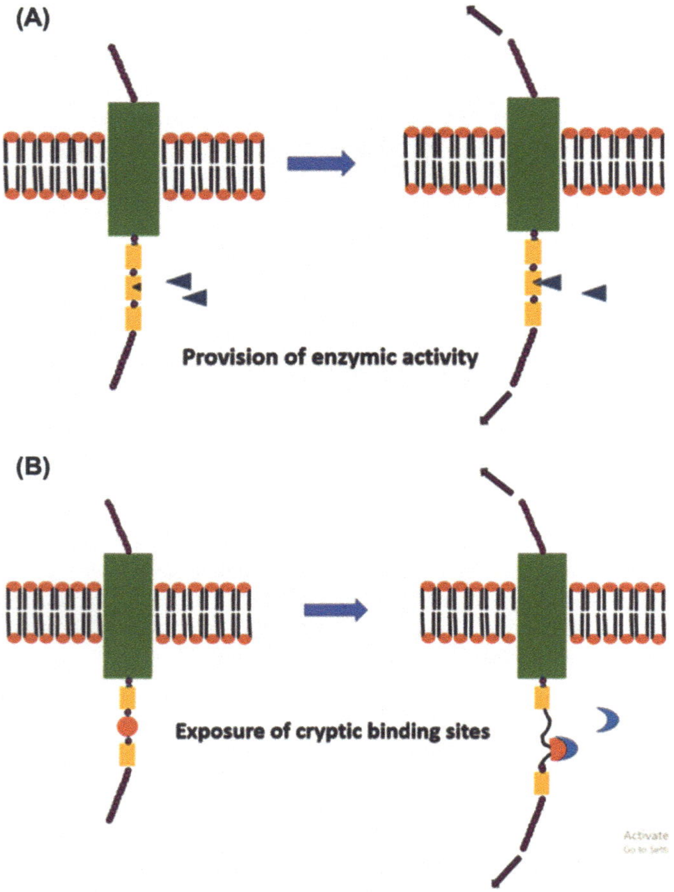

Figure 4.4 Conformal changes in proteins. Mechanical stimuli can trigger conformal changes in proteins to (A) facilitate binding or (B) expose new binding sites. *Orr A.W., Helmke B.P., Blackman B.R., Schwartz M.A., 2006. Mechanisms of mechanotransduction. Developmental Cell. 2006; 10, 11–20.*

are also expressed under intermittent mechanical strain. Mechanical stimulus also has other effects on intracellular processes. For example, in lung cells, stretching increases the activity of protein tyrosine kinases and receptor tyrosine kinases such as Flk-1 and stimulates the release of parathyroid hormone–related peptide in fetal lung epithelial cells, which is important for activating cyclic adenosine monophosphate (cAMP) to carry out specific functions of fibroblasts. Moreover, in fetal lung cells, the effect of applied strains promotes growth mediated by enhanced production of a number of growth factors including platelet-derived growth factor. Alveolar cells exposed to mechanical stimulus release inflammatory mediators such as interleukin-8 based on the magnitude of strain. Prostaglandin is downregulated in human airway epithelial cells when subjected to cyclic stretch caused by the inactivation of cyclooxygenase. However, prostacyclin production is enhanced by shear stress applied on lung endothelial cells.

4.3 Effect of mechanical stimuli on stem cells

Stem cells are simply defined as those cells with the capacity for self-renewal and conversion to other mature tissue. The types of stem cells widely used in research are the mesenchymal stem cells (MSCs). The relative ease with which these MSCs can be isolated has rendered them popular in stem cell research. In addition, these cells result in the formation or regeneration of multiple tissues including bone and cartilage. MSCs are easy to culture and have the capacity to differentiate into different lineages such a chondrocytes, osteocytes, and adipose tissue. MSCs have much promise for tissue engineering and regenerative medicine technologies. They are typically sourced from adult bone marrow as bone marrow stem cells (BMSCs) but can also be obtained from muscle, tendon, and adipose tissue. Biochemical drivers such as growth factors, although not necessary, are very useful for promoting the differentiation of MSCs to different lineages. For example, transforming growth factor β (TGF-β) is observed to favor chondrogenesis from MSCs, whereas bone morphogenetic proteins (BMPs) accomplish osteogenesis. The role of cytokines and growth factors in influencing the differentiation of MSCs is well known. However, these cells also have the tendency under specific circumstances to differentiate under mechanical stimulus that might be critical to the understanding of clinical treatment of load-bearing tissues. The influence of mechanical stimulus on stem cells is a fairly new area of study but has gained much impetus recently. Three differentiation pathways for MSCs have been well studied: osteogenic, chondrogenic, and adipogenic. However, stimulus magnitudes, type of force, duration of force application, and frequency are important parameters that affect stem cell differentiation, which have not been defined so far. This is attributed to the complex interaction between these parameters, and as a result, no regime of dominance of any of these differentiation lineages has been demarcated.

4.3.1 Incongruency in cell response

The ability of MSCs to differentiate into multiple cell lines under a variety of conditions implies that there is an innate difficulty in homogeneity of results in different studies, which then complicates the comparison of such results in an objective manner. Another viable option in stem cell research is to use pluripotent stem cells (PSCs), including induced PSCs and embryonic stem cells (ESCs), which have the capacity to differentiate into more well-defined but primitive cell lines, clearly of various cells in the body. However, the complicated conditions of in vitro cell culture required for these stem cells increases many-fold the difficulty to use them effectively and consistently in scientific studies. As a caveat, the variety in stem cell characteristics in the cell types extracted from various biological sources further complicates the analysis and comparison of results and prevents scientists from reaching universal conclusions in terms of stem cell behavior under mechanical effects. Thus the consistency in guiding stem cell fate remains a challenge, especially the ability to isolate the effect of mechanical stimulus from other underlying experimental factors. Hence, there is an urgent need for continued research in consistently guiding stem cell fate or even simply encouraging their proliferation with the proper combination of local and systemic factors. In the sections that follow, in view of the inherent difficulty in separating the effects of individual experimental parameters, the reader is warned that only a feeble attempt is made to isolate the effect of mechanical stimuli from other cofactors in describing the response of stem cells to external mechanical signals.

4.3.2 Types of mechanical stimuli

Mechanical cues have been reported to influence stem cell differentiation and proliferation. Similar to normal cells, some of the factors that affect stem cell behavior are substrate stiffness, nanotopography of the substrate surfaces, mechanical forces, and fluid flow even in the absence of biochemical factors. It is impossible to describe the effect of purely applied stress on stem cells because inadvertently other factors such as substrate/scaffold type, hydrogel coatings, cell culture medium, scaffold dimensionality, substrate topography, and chemistry play significant roles in affecting cell physiology and function. Even with these limitations, attempts have been made to maintain the combination of experimental factors to a minimum to study the response of stem cells to externally applied mechanical stimuli. It must be kept in mind that the direct influence of mechanical stimulus alone in isolation is difficult to accomplish, and a combination of factors appears to be more effective in achieving stem cell differentiation. The difficulty in isolating the effects of each of these parallel factors is succinctly illustrated in Fig. 4.5.

The myriad of factors involved in influencing stem cell fate includes mechanical, material, and temporal parameters. The complex and parallel actions of each of these factors demonstrate the difficulty in isolating and studying each parameter individually.

Many studies have reported the effects of different stresses on MSCs under a variety of conditions. The types of stresses can be divided into tensile (many times called stretch) and compressive stresses and also very importantly shear stresses specifically related to fluid flow. When referring to stretching, it must be emphasized that the stresses are applied to substrates on which the stem cells are cultured, and hence, cells are not directly subjected to tensile stresses. Usually, substrates with varying elasticities are used, and these are pliable enough to support large elastic strains up to the order of 40% and usually coated with cytocompatible hydrogels such as fibronectin or collagen, which mimic the extracellular matrix to facilitate cell adhesion and function. The applied strain is then transferred from

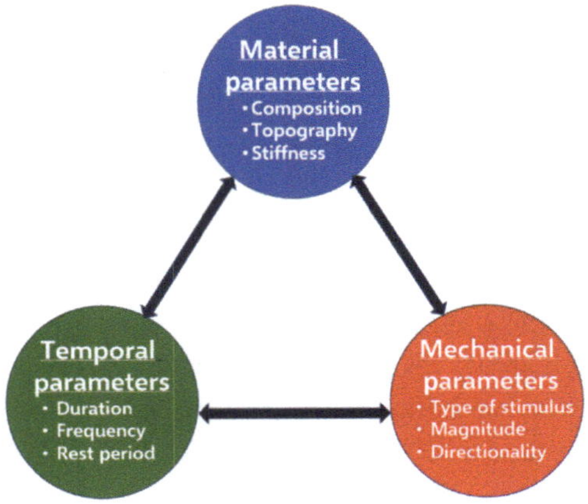

Figure 4.5 General factors affecting stem cell response.

the substrate to the cells via the ECM fibers. These are, in turn, connected to the focal adhesions, which are complex protein complexes, starting with the transmembranal integrins and connector proteins, such as paxillin, talin, vinculin, etc., and furthermore to the actin microfilaments. Actin filaments apply stresses to the nucleus at the end of the chain and bring about transcription and protein synthesis through mechanisms referred to in earlier chapters. Mechanical properties of the ECM also control stem cell fate.

Even so, scientists have highlighted the effects of stresses on MSCs as a significant factor in their proliferation and differentiation and critically related these to various embryogenetic processes in developmental biology. The direct effect of mechanical stimulus on MSC differentiation has been reported for different load types. In studies involving stretching forces, the effect of cyclic loads is sometimes included because in real physiological conditions, application of mechanical stimulus is in many cases intermittent and not constant. Very early studies on MSCs showed that an important driving force in their differentiation is cyclic uniaxial stretch.

4.3.3 Stem cell response to mechanical stimuli

4.3.3.1 Effect of stress type

The prime focus of stem cell response is their fate. Stem cell differentiation is accomplished in the laboratory through using soluble factors. Recently, mechanical stimulus has been touted as a parameter to accomplish selective differentiation, although the magnitude or the duration of force application for this purpose has not been defined clearly. Stem cell fate, as a result of differentiation, can also be effected as a result of stretch in MSCs. Stretching can be uniaxial or biaxial usually based on the stretching device used to study this effect. Considering the flexible substrates such as PDMS that are normally used, care must be taken in the analysis of the resulting data depending on whether the cells are subjected to uniquely uniaxial stretch or if transverse effects come into play during the test.

As in normal cells, the intermittency in stimulus application appears to be very important. Cyclic stretching alone has been shown to induce MSC differentiation, whereas human ESCs mature into cardiomyocytes under this stimulus. The type of cyclic stretch, whether uniaxial or biaxial, also has an effect on differentiation. Human bone marrow stromal cells under uniaxial stretch conditions expressed alpha-smooth muscle actin ($\alpha - SMA$), showing a tendency to SMCs, whereas biaxial stretch downregulated this gene. Biaxial stretch within low strain limits encouraged osteogenesis. For the cardiovascular system, cyclic stretch, cyclic hydrostatic pressure, and laminar shear stress are three types of mechanical stimuli, which are physiologically relevant in the vasculature. For MSCs under these three types of stimuli, size, shape, and alignment of these cells are altered. Cyclic stretch and cyclic hydrostatic pressure render MSCs significantly smaller in size, whereas under laminar shear stress, they grow larger. The response to cell shape appears to be dose dependent, and these become more spindle shaped at higher shear stress, although proliferation is preferred at laminar shear stress. In contrast, MSCs are more rounded when cyclic hydrostatic pressure is applied on the cells. The density of MSCs seeded on substrates are sensitive to cyclic stretch at low frequency, whereas at higher frequency, mechanical stimulus appears to have no effect whatsoever on cell density. Protein expression is strongly dependent on magnitude and frequency under cyclic stretch, as these stem cells differentiate into SMCs, whereas such a trend in cell fate is not observed under cyclic hydrostatic pressure or laminar shear stress. Cyclic stretch and compression of MSCs have been reported to cause osteogenic differentiation with increased expression of alkaline phosphatase (ALP). Even in 3D scaffolds designed for the fabrication of tissue constructs, the application of a tensile load stimulus on MSCs has been effective in the upregulation of BMPs implicating osteogenic tissue. In contrast, cyclic hydrostatic pressure and laminar shear stress implicate ECs-related gene expression pointing to a preference for these cells after differentiation. Even if only the SMC and EC lineages are preferred under cyclic hydrostatic pressure and laminar shear stress, small amounts of osteopontin mRNA have been detected, implicating atherosclerosis in vascular SMCs and ECs. Cyclic stretch also resulted in greater gene and protein expression in human ESC cardiomyocytes, although it appears that stretching alone is not sufficient to form mature cardiac tissue. Cyclic uniaxial stretching of endothelial progenitor cell (EPCs) increased VSMC marker α-SMA and downregulated the expression of Flk-1. Cyclic stretching also increases proliferation and encourages cell alignment perpendicular to the direction of applied stretch. The application of a dynamic tensile and torsional strains has resulted in guiding MSCs to a fibroblast lineage. In general, muscle-related genes are downregulated under cyclic hydrostatic pressure and laminar shear stress. A softer substrate causes chondrogenic rather than osteogenic differentiation under cyclic stretch, whereas a larger strain appears to favor the latter. Interestingly enough, chondrocyte differentiation has been implicated under applied pressures greater than 5 MPa in seeded MSCs, whereas osteogenic differentiation is supported at lower compressive strains. Cyclic compression and cyclic hydrostatic pressure increase the differentiation of BMSCs into the chondrogenic phenotype confirmed by the presence of collagen II, aggrecan, and other proteoglycans. As in the osteogenic case, chondrogenesis is further enhanced by the presence of TGF-β. Similar observations are made in 3D cultures, indicating that compressive stresses with a cyclic component play a critical role in the development of soft tissues, such as cartilage. In a general sense, tensile stretch favors osteogenesis and tenogenesis, and adipogenesis is more prone to compressive stress, whereas chondrogenesis results from hydrostatic pressure and compressive stress. Any type of tensile stimulus is counterproductive to the formation of adipogenic tissue.

The effect of shear stress on stem cell fate has been investigated even more than tensile and compressive stimuli, considering the ubiquity of fluid flow in different tissues. In one sense, it is easier to study shear stress effects on stem cells because of the relative simplicity of experimental design and geometry. Again, based on the magnitude and frequency of the applied shear stress, a variety of signaling pathways have been observed. One generally accepted fact is the dominant role of the primary cilium as a mechanosensor of the stimulus in the form of shear under fluid flow on a variety of stem cells. Any lack of osteogenic differentiation in stem cells under the influence of a shear stress has been attributed to the absence of primary cilia, and the presence of this mechanosensor has been clearly associated with the expression of osteogenic markers. When exposed to fluid shear stresses, stem cells tend to upregulate endothelial markers such as CD31 and cadherin, manifesting a tendency toward the formation of mature endothelial cells. EPCs tend to express the Flk-1 endothelial cell—specific marker under the influence of fluid flow shear stress. These also upregulate vascular endothelial growth factor receptors, intracellular adhesion protein, and vascular cell adhesion protein, consonant with cell—cell adhesion typical of endothelial cells.

MSCs also orient themselves and spread on substrates under applied loads in two dimensions, and substrate strain can induce alignment and elongation of these cells in the direction either parallel or perpendicular to that of the principal direction of tensile stretch. There is some general thinking that the cell alignment is a response to counteract the tensile force. The alignment of MSCs is parallel to fluid flow in bioreactors. MSCs preferred to align themselves rather than assume a random orientation under the action of applied cyclic stretch on a variety of substrates, although the orientation of MSCs perpendicular to the direction of applied stretch is sometimes observed. EPCs orient themselves parallel to the direction of fluid flow while their cell density and size increased. The discrepancy in these results is testimony to the complexity of this topic and the difficulty in maintaining uniformly similar conditions.

4.3.3.2 Effect of stress magnitude

Some semblance of relative quantification has been ascertained with regard to the differentiation of human and mouse BMSCs under various types of mechanical stimuli. Low strain magnitude of less than 3% supports osteogenesis, whereas strains greater than 10% favor tenogenesis. The 10% strain stimulus upregulates the tenogenic genes

Col I, Col III, and scleraxis, which is a tendon transcription factor. Application of cyclic strain at the 10% level increased Col I, Col III, and tenascin-C mRNA in rat BMSC culture, although at lower frequency, upregulation takes longer. It also appears that constant strain stimulus inhibits tenogenesis. It is hypothesized that myogenic differentiation of MSCs can be accomplished by mechanical stimulus in tension because this process is weak with the sole use of biochemical factors. Tension upregulates gene and protein expression, whereas neither Col mRNA nor ALP was detected in the MSC culture. A 10% tensile stretch at a 1 Hz frequency appears to be the optimal stimulus for cardiomyocyte gene upregulation. It must be noted that there are limits to the range of applied mechanical stimulus. Extremely low values of load or strain may not encourage proliferation, let alone enable stem cells to differentiate. On the other hand, highly intense stimuli may be detrimental to cell viability and result in apoptosis. Short spurts of applied stimuli are more beneficial in accomplishing stem cell differentiation than a sustained stimulus. Cyclic tensile loading is appropriate for the upregulation of early osteogenic markers and increases deposition of the mineralized phase. In addition, Col I mRNA and ALP activity are upregulated. Although the latter is observed solely under mechanical stimulation, dexamethasone medium was necessary for detecting early osteogenic markers such as osteopontin and osteocalcin. A strain magnitude of 10% in compression induced osteogenic genes in MSC, whereas both chondrogenic and osteogenic genes are induced by 15% compressive strain. Although the type of tissue formed is not clear, compressive mechanical stimulus gives rise to a nonfibrous cartilage-like tissue. There is a preconceived notion that because osteoblasts and osteocytes experience fluid flow stress in vivo, exposure of MSCs to fluid shear stress would lead to osteogenic differentiation. Although this appears to be true in experimental observations, fluid shear stress appears to have a stronger effect on matrix formation, which occurs at a later stage rather than the initiation of MSC differentiation to osteogenic tissue. Also, oscillatory fluid flow for short spurts appears to intensify matrix deposition.

4.3.3.3 Effect of substrate characteristics

The stiffness of the substrate on which stem cells are seeded also appears to play a role in their fate. Cell differentiation appears to be regulated by the stiffness of the ECM surrounding stem cells. MSCs can differentiate into neuroblasts, chondroblasts, or osteoblasts based on the stiffness of the gel in which they are seeded. Softer hydrogels also appear to favor differentiation of MSCs to adipogenic tissue, whereas a stiffer gel results in an osteogenic lineage. MSCs differentiated into different lineages as a result of culturing on substrates of increasing stiffness: 0.1−1 kPa, 20−25 kPa, and 30−35 kPa for neurogenic, chondrogenic, and osteogenic lineages, respectively. This stem cell fate guidance by durotaxis is also being generally accepted by the scientific community. Haptotaxis is another manner for the guidance of stem cell fate. Rounder and smaller cells appear to move toward adipogenic tissue, whereas a higher aspect ratio of the substrate topography definitely appears to guide MSCs to osteogenic differentiation. Although some controversy exists regarding the optimum size ranges for bias toward one or the other lineages, it is generally agreed on that grooves in the underlying substrate produce stem cell alignment and thus control the geometry of resulting tissues. Reports indicate a general acceptance of the combination of mechanical load and substrate nanoarchitecture in guiding stem cell fate even in the absence of chemical cues. Stem cell fate is also reported to be affected by the duration of the applied mechanical stimulus on a specific substrate. A softer substrate results in the expression of osteogenic markers if the duration of culture is short, whereas a longer dwell time on the softer substrate results in adipogenic tissue.

4.3.3.4 Effect on extracellular matrix

Another important result of the effect of mechanical stimulus on MSCs is the increase in the deposition of ECM. Studies indicate that matrix synthesis is enhanced significantly in MSCs by mechanical loads, regardless of substrate architecture or its fiber alignment. This is evidenced by the increased synthesis of collagen Type III, which is also an indication of MSC differentiation into fibroblasts. However, as mentioned earlier, neither mechanical stimulus nor substrate topography alone is a sufficient condition for MSC differentiation into fibroblasts. MSCs seem to interact with the ECM through the integrins indicating that these play a critical role as mechanosensors. It is clear from the plethora of data that integrins are upregulated in MSC cultures under the influence of mechanical signals again pointing to these transmembranal structures as strong mediators in the process of mechanotransduction.

Although the general acceptance is that mechanical stimulus alone is not sufficient to determine stem cell fate, there is another emerging school of thought that mechanical stimulus alone may be sufficient to provoke stem cell differentiation. For example, osteogenic differentiation manifested by the presence of CBfa1/Runx2, osteopontin, and osteocalcin can be induced by mechanical stimulus alone. Also, differentiation into osteogenic tissue can be accomplished even without osteogenic supplements by the use of an appropriate mechanical stimulus. It is argued that biochemical factors including special media and the presence of other supplementary factors although not necessary can help enhance the MSC differentiation process. As stem cell differentiation in vivo does involve multiple factors, the use of combined parameters in tissue regeneration may be more effective and faster in driving cell fate. As an example, the differentiation of MSCs into fibroblasts requires costimulation with an aligned nanofiber substrate (biomimetic structure) and mechanical loading as manifested by collagen Type III synthesis. The resulting matrix is a more ligament-like material deposited by such fibroblasts based on interactive costimulation. The fibroblasts orient themselves in the direction of stretch but remain unaffected in alignment by the applied tensile load. Aligned scaffolds, as another coupling factor, also encourage a greater cell density and a significantly enhanced total matrix deposition under a tensile stimulus. Currently, there are varied opinions as to which mechanical stimulus parameters and their corresponding magnitudes are effective in stem cell fate guidance. For the present time, it is enough to indicate that mechanical stimulus is indeed an important factor both in the laboratory and in vivo to achieve stem cell differentiation.

Table 4.1 summarizes the results gathered from a number of scientific studies and is included with the objective of providing a perspective into the varied responses of mesenchymal stem cell lineages to different mechanical stimuli. As mentioned earlier, it must be emphasized that it is difficult to obtain the same response from a particular type of stem cell even if the experimental conditions and the applied stimulus is the same.

Although much is unknown at present, the properties of the ECM may regulate stem cell biology. Just as in normal cells, stem cells are also sensitive to the ECM-integrin machinery for determining fate. As stem cell fate is decided by stiffness variations of the ECM, where the more mechanically robust tissues are generated from stem cells when ECM stiffness is higher, it is believed that the RhoA/ROCK activity ultimately leads these cells to an osteogenic commitment. At least, in vitro, depending on the biomechanical characteristics of the substrate on which stem cells are grown, they tend to differentiate into various cell types by expressing markers appropriate to these cells. Polymer gel substrates tend to favor neurons, whereas with increasing stiffness, skeletal muscle tissues or even markers associated with osseous tissue may be expressed by MSCs. Thus regenerative medicine holds the hope that by controlling the biomechanical properties of the ECM in stem cell niches, desired tissues may be generated by the modulation of MSCs.

4.4 Effect of mechanical stimuli on cancer cells

It is obvious from the earlier sections that mechanical stimulus influences cell proliferation and differentiation in normal and stem cells. However, not much is understood about the effect of such stimuli on cancer cells, although our knowledge in this area has been growing recently. The cell cycle is a critical physiological phenomenon that is characteristic of nonhematopoietic cells. To accomplish this, adherent cells require a substrate to attach to, spread, and progress through this cycle. Cells, which either lose this ability through mutation or are simply unable to attach and complete the cell cycle normally, are very similar to tumorigenic cells in vivo, which have the capacity to migrate, invade, and metastasize. The adherence of cells on a substrate is consonant with a surface with appropriate stiffness. Cells attach to a surface with a certain contractile stress or cell traction force (CTF), measured at 35.5 nN/μm^2. Transformed cells on the other hand have a CTF of approximately 1 nN/μm^2. Compliant substrate surfaces therefore inhibit normal adherent cells from attaching and spreading effectively, while such surfaces with relatively lower stiffness are attractive to tumor cells.

Table 4.1 A useful compendium of the wide variety of responses of various human stem cells to different types of mechanical stimulus: substrate stiffness, substrate topography, and type of applied stress.

Stem cell type	Type of stimulus	Conditions	Response	Cell lineage commitment
MuSCs	Substrate elasticity of PEG hydrogels	2–42 kPa stiffness, 2–7 days	Cell proliferation occurs, lower cell mortality rate	Stemness retained
MSCs	Substrate elasticity of polyacrylamide gels	0.1–40 kPa, 1–3 weeks	Changes in morphology corresponding to neurons, myoblasts, and osteoblasts based on gel stiffness	Neurogenic, myogenic, osteogenic
MSCs	Substrate elasticity of PDMS	1.5 kPa and 50 kPa	Normal cell proliferation, slight tendency in differentiation to adipogenic or osteogenic lineage	Adipogenic or osteogenic
MSCs, iMSCs	Substrate elasticity of hPL gels	20–30 Pa	Normal cell proliferation and differentiation	No lineage specific differentiation
Stem cell type	Type of stimulus	Conditions	Response	Cell lineage commitment
MSCs	Geometric cues	Different aspect ratios of 1:1, 3:2, and 4:1, 1 week	Geometric features, which enhance actomyosin contractility are osteogenic, otherwise adipogenic	Adipogenic or osteogenic
MSCs	Fibrin islands	Different areas, from 1024 to 10,000 μm^2, 1 week	Cell shape controls stem cell fate, adipogenic if rounded and osteogenic if spread out.	Adipogenic or osteogenic
MSCs	Concave and convex curvatures	Spherical structures with 250–750 μm diameters, 1–10 days	Significant increase in cell mobility on concave surfaces. Substantial nuclear deformation and lamin-A expression on convex surfaces	Osteogenic

(*Continued*)

Table 4.1 (Continued)

Stem cell type	Type of stimulus	Conditions	Response	Cell lineage commitment
MSCs	Vertically aligned silicon nanowires	10–25 nm length and 150–200 nm diameter, 1–3 days	COL1A1, Runx2 F-actin, pFAK, vinculin, and alpha 2 integrin expression on shortest nanowires significantly higher. Nanowires regulate osteogenic differentiation.	Osteogenic
MSCs	TiO_2 nanotubes	30–100 nm, 3 weeks	Higher cell contractility on larger nanotubes causing osteogenic differentiation.	Osteogenic
MSCs	Stretching and torsion	Translational 10%, rotational 25% in collagen gels	Increased expression of collagen Types I and III, tenascin-C, increased cell alignment, density, formation of oriented collagen fibers	Ligaments
MSCs	Cyclic uniaxial stretching	10% and 12%, 4 h/day for 7 and 14 days in collagen matrices	Increase in BMP-2 mRNA levels	Osteogenic
ESC-CMs	Cyclic uniaxial stretching	12%, 1.25 Hz, 3 days	Greater proportion of cardiac troponin T-expressing cells, higher proportion of cardiomyocytes, more mature phenotype, enhanced mRNA expression of genes	Mature cardiomyocytes
BMSCs	Cyclic uniaxial stretching	8%, 3 days	Increase in the gene expression of Bfa1, ALP, OCN, collagen Types I and III	Osteoblasts

(*Continued*)

Table 4.1 (Continued)

Stem cell type	Type of stimulus	Conditions	Response	Cell lineage commitment
ESC-CMs	Steady or cyclic uniaxial stretching	10%, 1 Hz, 3–14 days	More mature cardiomyocytes in terms of twitch force, elastic modulus, sarcomere length and molecular signature, when comparing to static stretch or nonstretched controls.	Mature cardiomyocytes
MSCs	Cyclic uniaxial stretching	5%, 1 Hz, 2–4 days on collagen I—coated micropatterned silicone	Increase in gene expression of SMC marker calponin 1; decrease in gene expression of cartilage matrix markers (biglycan, COMP, collagen type X1a, and collagen type XI a1	Smooth muscle
BMSCs	Cyclic uniaxial and equibiaxial stretching	10%, 1 Hz for 1–3 days on collagen or elastin membranes	Transient increase in α-SMA, SM 22a, and collagen Type I gene expression by uniaxial stretching downregulation of SM α-actin and SM-22a by equibiaxial strain	Smooth muscle
MSCs	Cyclic equibiaxial stretching	3%, 8 h	Increase in gene expression of cBfa1, OCN, and ALP	Osteoblasts at low magnitude (3%), short duration (8 h)
		8%, 48 h	Increase in gene expression of collagen types	Tendon/ligament at high magnitude (8%),
BMSCs	Cyclic compression	7994 Pa, 0.33 Hz, 7 days	Increase in the gene expression of collagen Type II and aggrecan	Chondrogenic
MSCs	Cyclic compression	10%, 0.1, 0.5, and 1 Hz, 8 h/day, 2 days in fibrin gels	Increase in aggrecan gene expression at all frequencies; increase in collagen Type II at 1 Hz only	Chondrogenic

(*Continued*)

Table 4.1 (Continued)

Stem cell type	Type of stimulus	Conditions	Response	Cell lineage commitment
MSCs	Cyclic uniaxial compression	0.22%–1.10% strain, 1 Hz	Increased expressions of OCN and COL1A1, 3.76–fold increase ALP activity.	Osteogenic
MSCs	Intermittent hydrostatic pressure	0.1 MPa, 14 days	Increase in gene expression of Sox-9 and aggrecan	Chondrogenic
		10 MPa, 14 days	Increase in collagen Type II gene expression	
MSCs	Intermittent hydrostatic pressure	1 MPa, 1 Hz 4 h/day, 10 days in collagen sponges	Increase in gene expression of collagen Type II, aggrecan, Sox-9, and collagen Type I; increased accumulation of proteoglycan; no change in Runx2 mRNA levels	Chondrogenic
MSCs	Cyclic hydrostatic pressure	Steady at 7.5 MPa for 14 days, 4 h/day at 1 Hz in agarose gels	Transient increase in Sox-9 gene expression by both types of loading	Beginning stage of chondrogenesis
		Ramped at 1 MPa at day 1, 0.5 MPa increase in subsequent days, 7.5 MPa at day 14, 4 h/day at 1 Hz	No change in collagen type II, and aggrecan gene expression	
MSCs	Shear stress	Oscillatory pressure driven flow, 1–5 Pa, 0.5–2 Hz, 2–4 h	2 Pa, 2 Hz upregulates osteogenic gene expression. Significant increase in collagen and mineral deposition.	Osteogenic

ALP, Alkaline phosphatase; *BMP-2*, bone morphogenetic protein-2; *BMSC*, bone marrow stem cell; *CM*, cardiomyocyte; *EMC*, embryonic stem cell; *MSC*, mesenchymal stem cell; *MuSC*, muscle stem cell; *OCN*, osteocalcin; *PDGFR*, platelet-derived growth factor receptor; *PECAM-1*, platelet endothelial cell adhesion molecule-1; *PEG*, polyethylene glycol; *SMC*, smooth muscle cell; *TGF-b*, transforming growth factor-b; *VEGFR-2*, vascular endothelial growth factor receptor-2; *VSMC*, vascular smooth muscle cell.

Contrarily, tumor tissue has higher stiffness, and this may affect its architecture. Instead of hollow cysts that form in normal tissue, stiffer tumor tissue encourages the transformation of its structure to fill cystic lumens and disrupt cell polarity as observed in breast carcinoma. When the actual routing of the mechanical stimulus sensed through integrin clusters and transmitted to the nucleus through translocation is disrupted, cells lose their ability to polymerize actin. Inhibitors of actin and myosin polymerization, in turn, diminish the ability of growth factors to sustain MAPK activity. This affects the expression of Cyclin D1, which is critical for cell cycle progression in anchorage-dependent cells.

The tumor microenvironment is dynamic and mechanically sensitive to the continuous changes taking place in the tissue. The mechanical properties of this microenvironment are constantly changing with continued remodeling of the ECM, resulting in various stresses. As the ECM stiffness increases, contractile stresses are enhanced in cells in the tumor microenvironment. The proliferation of these cells in a confined space imposes an increasingly greater compressive stress, which may cause apoptosis. Tumor growth also increases blood and interstitial fluid flow resulting in an applied shear stress on the cells in the microenvironment. As it is continuously remodeled, the interaction between the cells and the alterations in the microenvironment may drive these cells to malignancy. Abnormal ECM dynamics in the cancer cell niche and the metastatic niche may be another factor, which strongly influences the progressive stages of cancer. Both decreased production of certain ECM molecules and the unregulated excess of others in the ECM may cause tissue fibrosis leading to tumorigenesis in the vicinity. Increased tissue stiffness resulting from a high collagen content has been known to encourage tumor formation. Tumor stroma is considerably stiffer than normal stroma, and progressive increase in the stiffness of sensitive tissue can be a definitive sign of cancerous tissue development. The higher stiffness in tumors is attributed to the deposition of mainly Type I collagen and fibronectin. Integrins have been strongly implicated in tumor progression because increased stiffness in the tumor microenvironment upregulates integrin signaling. This, in turn, promotes integrin clustering and encourages cell migration to premetastatic niches. The transport routes laid by oriented collagen fibers in the tumor region are responsible for invasion and intravasation. Increased integrin signaling also enables cell survival and proliferation.

The deregulation in ECM dynamics is another factor that also affects normal mechanotransduction, which influences cell differentiation and tissue polarity, leading the way to cancer progression. The deregulation of ECM remodeling may result in various human cancers. Excess lysyl oxidase (LOX) encourages the cross-linking of collagen fibers and hence increases the stiffness of the ECM as has been observed in various carcinomas. This has been suggested to play a strong role in promoting tumor cell invasion, indicating that this increased stiffness of the tissue may be a cause of cancer. Also, compressive stress in the tumor microenvironment is a likely culprit in the selection of cells for metastasis and invasion. A low level of compressive stress increases cell motility and probably aids in intravasation and extravasation, but excessive levels trigger apoptosis. Cofilin has been implicated in sensing mechanical signals for enhanced invasion during metastasis. Much needs to be still learned and understood about the effect of mechanical signals on cancer and tumor cells.

4.5 The role of cell adhesion molecules in mechanosensing

In contrast to cells with enhanced motility, adherent cells respond in different manners to applied mechanical stimulus. The transmission of this signal to elicit a cell response occurs through the connections between the substrate, which provides the first-level reaction to the applied force or deformation, and the cell nucleus, which is the receptor of this signal to bring about gene transcription and protein synthesis. The intermediaries that serve as the conduit for this signal are first, the ECM molecules which are then connected to the transmembranal heterodimeric integrins and further on to the actin stress fibers via complex focal adhesion proteins. The mechanosensory role has been exclusively assigned to the integrins, although it is believed that the complete FAC must be almost simultaneously nucleated and activated for appropriate sensing of the mechanical signal. Other factors that are critical to this process include the substrate/ECM stiffness, the topography of the substrate, ligand spacing, and the temporal and spatial characteristics of the applied signal.

Cells do not adhere adequately to typical substrates that are normally used to culture them in vitro. Substrates generally used are PDMS, poly-lactic acid, polyethylene glycol, and tissue culture plastic, which is a modified polystyrene. These artificial substrates facilitate the application of a variety of mechanical stimuli. However, for effective cell adherence, these substrates have to be coated with ECM proteins, such as fibronectin, collagen, or short Arg-Gly-Asp (RGD) peptides. These protein fibers then provide the necessary anchorage for cells through bonding with integrin heterodimers.

Although integrins are the first responders in the line of mechanosensing in the ECM, the nucleation and development of focal adhesions, which connect intracellularly to the integrins, are critical for the signal transmission process. The development of focal adhesions does not depend on a single factor but the ECM, integrins, and actin, all contribute to their nucleation and growth. It is believed that the formation of a focal adhesion is a result of a mechanosensory process as the cell approaches a substrate to attach to it. The mechanical properties of the substrate play an important role in this attachment process. Cells obtain a feel for the substrate through contractions, which they apply periodically to sense the stiffness of the substrates. These contractions generate contractile stresses whose magnitudes depend on the elastic modulus or stiffness of the substrate. If the substrate is too soft, a cell will not linger long on the surface and attempt to move to a stiffer region. Cell contractile forces may be high for the localized soft surface, which may wrinkle and destabilize the cell from attaching. However, an adequate stretch stimulus applied to the substrate may counteract the contractile forces of the cell, and if these are overcome, the cell will be encouraged to spread on the surface and attach effectively. Now, whether this process is based on physical energy—based principles or purely biological processes is debatable. The contractile forces then recruit integrin proteins to bond with the fibers in the laminin layer which the cell lays out. Of course, the bonding is preferential based on the type of fiber protein available in the substrate/ECM. This is one of the reasons for the necessity in using a cytocompatible coating on common polymeric substrates to encourage cell attachment. As stiffer substrates do not wrinkle easily in comparison to a softer substrate, cells prefer such a surface to spread out and adhere. This adhesion clearly occurs with the transmembranal integrins through intracellular FACs. At the other end, the type and number of integrin proteins that bond to the fiber proteins in the ECM depends on the attaching cell line.

Instigated by a mechanical stimulus, interaction of the cell with the ECM is initiated by binding with integrin receptors. The first set of focal adhesions called nascent adhesions are initially formed, and these then recruit adaptor proteins in the cytosol to tether to the cytoskeletal proteins, in particular the actin microfilaments. These then interlace with the thicker myosin II fibers via other signaling pathways and form the stronger stress fibers. Although it is well known that the nascent focal adhesions assemble and disassemble, such mechanisms are still unclear. All that is known at present is that an applied mechanical stimulus causes further growth of these adhesions, which are then called mature focal adhesion. In addition, if the substrate is stiffer, cells are able to exert higher traction forces on the substrate, which results in the formation of more and larger focal adhesions accompanied by cell spreading. A similar effect is observed when a stable mechanical stimulus is externally applied within reasonable limits.

Although it is clear that integrins nucleate individually, clustering of these proteins occurs as the cells spread on the substrate or when a mechanical stimulus is imposed on the cell. Thus indirectly, the application of mechanical stimuli causes remodeling of the ECM where fiber proteins in this matrix are recruited and grouped as a result of integrin clustering. Integrins are of course the main components of focal adhesions, but their internal connection to the cell is accomplished by a large group of proteins. Of these, a few are well studied and critical to the cell attachment process. Integrins are extremely important in the mechanosensory process and are constantly interacting and exchanging cues with the ECM to bring about adhesion, cell motility, and cell division. As the nascent focal adhesions mature, they cluster together as a result of integrin agglomeration and form FACs, which are reported to comprise of anywhere between 50 to 150 proteins. Fig. 4.6 is a clear but simple illustration of a focal adhesion and highlights only the prominent proteins in the complex.

Among other proteins, talin is a connector protein that tethers β-integrins to the actin microfilament. The application of mechanical stimuli appears to upregulate talin through an increase in cytoskeletal tension. Many other proteins, including paxillin, zyxin, vinculin, Src, p130Cas, and FAK, are recruited during the maturing process of

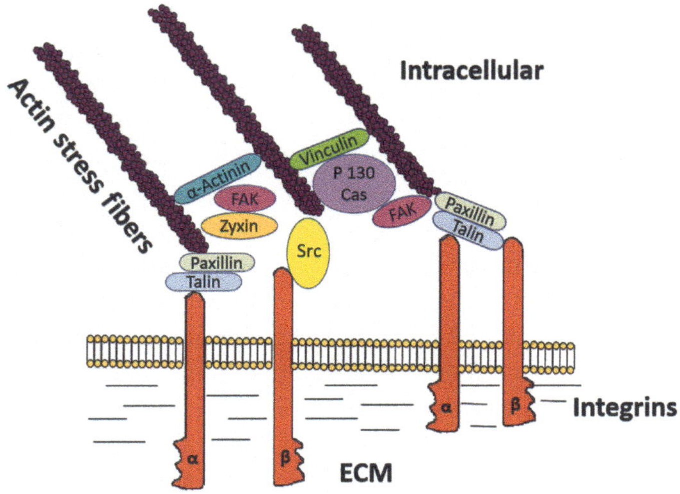

Figure 4.6 Focal adhesion complex. Schematic diagram of the focal adhesion complex (FAC) indicating the prominent proteins, which constitute the FAC. Connection of transmembrane integrins to the ECM and intracellularly through the FAC to the actin stress fibers is clearly exhibited. *ECM*, Extracellular matrix; *FAK*, Focal adhesion kinase.

the focal adhesions. Vinculin, in particular, appears to be in globular form in the cytosol and is activated by cell contractile stress or an external mechanical stimulus. On activation, the vinculin head attaches to the FAC, whereas its tail is attached to the actin stress fibers to act as an adaptor protein.

These FACs are clearly regulated by mechanical forces, whether internal or external. They must disassemble and reassemble during cell migration, which implicates their role in the shortening and lengthening of the actin stress fibers through the process of polymerization. In cells, some actin stress fibers extend from one focal adhesion to another, whereas others are connected to the cytoskeletal machinery. Thus there exists a feedback loop based on the sensory proteins, the assembly and disassembly of the focal adhesions, and the polymerization and depolymerization of the actin stress fibers through a complex interplay of signaling mechanisms. Fig. 4.7 summarizes this complex process through a series of simple steps.

4.6 Cellular proliferation

The influence of mechanical stimulus increases proliferation in both normal and stem cells. Even as the effect of mechanical stimuli on cell proliferation is described in detail, cell apoptosis must also be simultaneously considered because both these processes affect tissue homeostasis. Cell proliferation is described as the process when one mother cell divided into two daughter cells. This process of cell division consists of various phases including (1) the primary growth (G1) phase where a considerable amount of biological material is

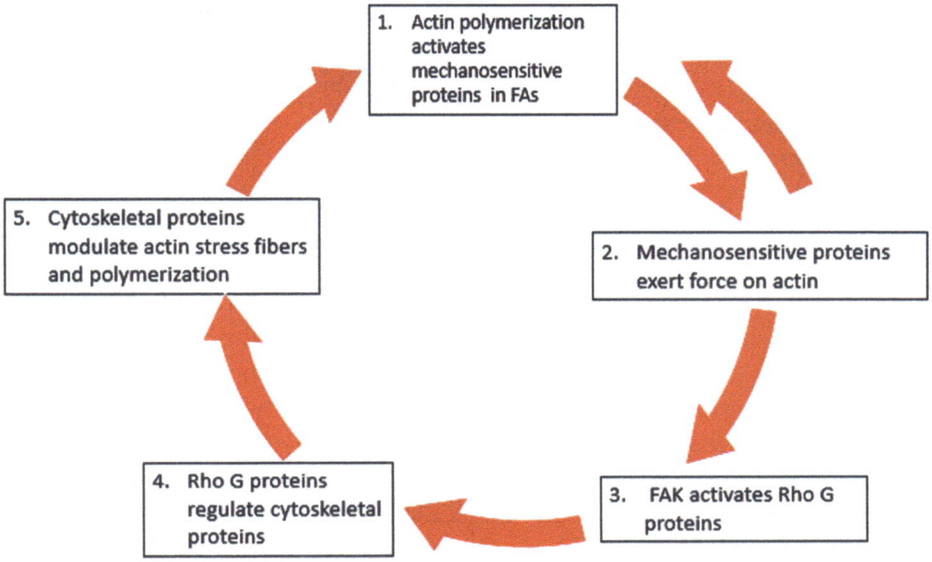

Figure 4.7 Feedback loop in mechanotransduction. The diagram depicts the feedback loop mechanism involved in mechanotransduction. Actin polymerization activates the mechanosensitive proteins in the focal adhesion complex (Step 1), which is reciprocated by mechanosensory machinery exerting a force on the cytoskeleton (Step 2). The signaling proteins such as focal adhesion kinase activate Rho G proteins (Step 3), which regulate cytoskeletal proteins (Step 4). These are tasked with modulating the actin stress fibers (Step 5) in a feedback process involving mechanotransduction. *ECM*, Extracellular matrix; *FAK*, focal adhesion kinase.

synthesized by the cell, (2) the synthesis (S) phase during which the cell replicates its DNA, (3) the second growth (G2) phase where further cell growth occurs, and finally (4) the mitosis (M) phase where the chromosomes are reorganized before actual cell division. It is important to point out for the sake of completeness that some cells enter the quiescent (G0) phase and temporarily stop proliferation. Although mechanical cues affect proliferation, the specific instants at which these signals act must be tightly coordinated with the biological maturity of the cell and the particular physiological timing to accomplish proliferation. Any shortcuts in this process will give rise to mutations and pathological defects. Matrix stiffness affects the proliferation capability of adult cells. In the case of muscle stem cells, cultures on substrates of intermediate stiffness appear to show proliferation potential and not on stiffer substrates. Another interesting aspect of in vitro proliferation is that regardless of the initial location of a mother cell, it migrates to a region on the substrate where it experiences a lower degree of internal deformation in the presence of smaller contractile stresses and hence greater stability before beginning mitosis. A higher substrate stiffness clearly contributes to this increased stability by decreasing the magnitude of cell contractile stress. Cell proliferation can also be initiated by cell-substrate interaction during cell migration. It is reported that cell matures faster on stiffer substrates because they suffer little internal deformation on these surfaces, and hence, their cell cycle is decreased, leading to increased proliferation. Stiffer substrates also contribute directly to increased cell growth. In this case, as in normal cells, the magnitude and frequency of the mechanical signals also affect the proliferation of stem cells. Excessive deformation of the cell, for example, by continuous tensile cyclic stretch of 10% and more results in apoptosis. Neuronal cell death is caused beyond 20% cyclic strain.

There is growing evidence that mechanical stress affects cell proliferation. This is rooted in the fact the cell division process somehow involves a change in cell shape. If an alteration in cell shape is accompanied by a remodeling of the cytoskeleton, any factor that affects the cytoskeleton will inevitably influence the changes in cell shape resulting from cell division. The cytoskeleton is easily the most important target for mechanical interventions, primarily because most stress on the cell is transmitted through the cytoskeletal elements or absorbed by these. By manipulating the cytoskeleton, one is able to change cell geometry and hence control cell behavior, including proliferation. Two related factors must be considered to understand this concept. First, when cell shape is modified as a result of an applied mechanical signal, cell progression is affected and, consequently, cell proliferation. Second, any aberration in alteration of cell shape, which compromises the structural integrity of the cell, can be a signal to initiate apoptotic processes.

Dynamic cyclic stretching of cells causes proliferation when compared with a static stretch. This has been indirectly linked to the formation of microtubule-organizing centers and the enhancement of polymerization of the microtubules under the action of a stress, which eventually triggers cell division. Both the magnitude of the applied stimulus and the frequency of the signal influence the process of cell proliferation. An optimum value of applied stimulus (within physiological limits) and the adequate frequency of application is most effective in enhancing cell proliferation in cell cultures. However, if the frequency of the applied stimulus is too low (<0.5 Hz) or too high (>1 Hz), cell proliferation is negatively affected. Similarly, if the applied stimulus is quite low in magnitude, it appears that cells do not respond, whereas a high magnitude stimulus results in apoptosis. There appears to be a dose-dependent relationship between the applied stimulus and proliferation of cells. It can be suggested that the cell cycle is shortened as a result of the applied stimulus, which is manifested as an increase in proliferation. It is also not well understood if it is deformation or the applied stress that is the cause of the enhanced cell proliferation under the applied mechanical stimulus.

4.7 Extracellular matrix remodeling

All tissues and organs have two main components: cellular and noncellular. The former provides biological definition to the tissue, whereas the latter in many ways ascribes structural, mechanical, and biochemical properties to the tissue. Such properties are important for the physiological behavior of the tissue and critical for carrying out cell function. This noncellular component is called the ECM and is composed of proteins, polysaccharides, and water. During tissue growth in the processes of embryogenesis and development, the ECM is constantly remodeled based on structural and biochemical requirements to maintain tissue homeostasis or advance tissue physiological function. Although the basic components are similar, each tissue-specific ECM has its distinct composition, topology, and hence specific biochemical and mechanical properties. The ECM also plays a critical role in mechanosensing and signal transduction processes, which are contingent to the dynamic evolution and reciprocity where cellular processes and the ECM interact to accomplish proper functioning of the tissue. ECM material is found in the basal membrane and in the intercellular space. The components of the ECM are secreted into the interstitial space after intracellular synthesis with the purpose of surrounding and protecting cells. The ECM is a dynamic regulator of cellular behavior and has many functions. It does not merely serve as a scaffold on which cells attach but is a surface on which these cells exert a traction force, which results in the modeling and remodeling of the ECM. There is a dynamic force equilibrium between the attached cells and the ECM, which helps in maintaining tissue homeostasis. Perturbation of this equilibrium can be exploited, and appropriate manipulation can encourage tissue regeneration. A few observations are mentioned here with regard to the mechanical properties of the ECM. MSC differentiation is directed by the elastic stiffness of the ECM, and the varying stiffnesses may be the driving force for cell migration. Moreover, the rate of cell motility is regulated and directed by the topography and geometry of the ECM. The modeling and remodeling of the ECM is a dynamic process, which responds to the requirements of the cell, whether sensing, anchorage, proliferation, migration, or any other cell function.

Two classes of macromolecules make up the ECM: proteoglycans and fibrous proteins. The proteoglycans are mainly composed of GAGs attached to a protein core, referred to collectively as proteoglycans. Proteoglycans are extremely hydrophilic, and their open structure enables them to absorb a large amount of water, which when combined with their branched conformations endows them with their hydrogel properties to withstand relatively large loads in compression. Proteoglycans have many roles in the ECM, including enabling cell surface receptors as well as controlling cell adhesion and migration. Fibrous proteins, on the other hand, provide the binding sites for anchoring during cell adhesion, contribute to the tensile strength of the tissue, and support cell migration. The most abundant fibrous protein in the ECM is collagen. Fibroblasts are the key source of collagen in the cell interstices and package the tropocollagen into fibers, which are oriented and aligned by the fibroblasts according to the need of the respective tissue. Elastin is another fibrous protein, which is a component of tissues that suffer repetitive stretching such as the skin. Fibronectin is another important fibrous protein, which directs the organization of the interstitial ECM. As this protein is prone to a large amount of stretching, it can uncoil and expose cryptic binding sites to which integrins can bind effectively. Thus fibronectin is an excellent mechanosensor and is hence able to regulate applied mechanical stimuli and assemble integrins implicating its role in cell adhesion and migration. The different ECM components in specific tissue define their physiological function and behavior. In cartilage and tendons, chondroitin sulfate provides support to these tissues, whereas SPARC (also known as osteonectin), an ECM-associated glycoprotein that binds cells and ECM components, is important in bone mineralization.

It is important to understand that the ECM is being constantly remodeled to carry out precise tissue function. This remodeling process involves two important steps to maintain tissue homeostasis: synthesis of the ECM components carried out intracellularly and their proteolytic breakdown. MMPs and their corresponding inhibitors work in a coordinated manner to maintain tissue homeostasis through balance of cell growth and apoptosis through various growth factors present in the ECM. A variety of cell—ECM interactions modulate the responses of tissues for the purposes of homeostasis, wound healing, or even disease progression. In the context of mechanotransduction, cells remodel the ECM when they are subjected to mechanical stimuli. There are six possible mechanisms by which cells remodel the ECM. These are briefly described below:

(a) Adhesion: Integrins are the principal transmembranal receptors that mediate cell adhesion to the ECM. By adhering to the ECM, cells not only transduce mechanical signals but also serve as sensors, which transmit extracellular information to the cell nucleus for gene expression and protein synthesis.

(b) Contraction: CTFs and contractile stresses serve to remodel the ECM by orienting and aligning the fibrous proteins present. Such a realignment serves to transmit signals over relatively large distances compared with cell dimensions. Besides, the alignment of collagen and fibronectin increases the stiffness and strength of the ECM and hence the mechanical properties of the tissue. Collagen alignment also provides tracks for directing cell migration along these fibrous proteins by contact guidance and hence reduces the energy expended by cells by decreasing drag.

(c) Degradation: Another important cellular mechanism contributing to remodeling is the degradation of ECM components by proteases. By secretion of these enzymes, cells are able to break down the proteins in the ECM to help repair damaged tissue during wound healing. It is important to mention MMPs, which is a special group of proteins, which are critical to cell physiology and play a critical role in developmental embryogenesis and wound healing. Diseases progression takes place when there is imbalance in the function of these degradants. MMPs cleave proteins in the ECM and remodel it in order for cells to migrate.

(d) Internalization: The degradation of the ECM by proteases occurs concurrently with endocytosis and lysosomal degradation. These ECM components are then internalized by the cell through endocytosis and later broken down by lysosomes. One important process is the internalization of fibronectin, which is critical for soluble collagen endocytosis. Thus both MMP degradation of the ECM and lysosomal activity must be balanced for tissue homeostasis.

(e) Deposition: Balanced deposition and degradation are important for a healthy tissue. Of the many cell types that contribute to the deposition of ECM components, fibroblasts are the major depositors of protein in the ECM, including collagen, fibronectin, and proteoglycans. It is believed that CTFs aid in the enhanced formation of fibrils of collagen by exposing cryptic binding sites on the ECM proteins to allow collagen and fibronectin to deposit. It is now known that the interaction between collagen and fibronectin is key to the formation of collagen fibers.

(f) Cross-linking: The stiffness and strength of the ECM are increased by crosslinking of the ECM proteins. The enzyme LOX is implicated in this process through covalent cross-linking.

Fig. 4.8 summarizes these six mechanisms, which are critical in ECM remodeling.

The constitution of the ECM and hence its mechanical properties are extremely important in its response to any applied mechanical stimulus. The molecules that comprise the ECM are functionally diverse, able to accomplish a large number of cellular activities, and constantly being remodeled, which provides a versatile platform for cells to continuously react to different changes in their environment with relative ease. Based on its composition and topography, the mechanical properties of the ECM vary widely and also changes constantly in a particular location. It can be soft and compliant or stiff and rigid, which influences the resulting cell and tissue behavior. Relatively small changes in the ECM stiffness can affect cell behavior. A change in collagen concentration in the ECM will increase its stiffness and cause clustering of the focal adhesions by driving the assembly of integrins. As mentioned previously, cell growth, activation of focal adhesion proteins, and cell—cell cadherin junctions are all affected by ECM remodeling. An increase in ECM stiffness can also cause the transformation of cells without the need for biochemical cues. It is also important to highlight the downside of ECM remodeling in disease progression, for example, in cancer. Overexpression of MMPs in cells followed by the rapid degradation of the ECM and realignment of collagen fibers may lead to cancer metastasis. The stiffness of a tissue or a tumor is an indicator of disease progression because tumors have a different stiffness from that of healthy tissue. In fact, an excessive increase in stiffness leads to tissue dysfunction and disease progression. The higher stiffness of the arterial wall is an accepted indicator of atherosclerosis. Changes in stiffness in the vicinity of an implant have also been known to cause tissue fibrosis and tumor growth. Furthermore, the excessive remodeling of the ECM increases tissue stiffness and facilitates cancer cell invasion to the adjacent stromal tissue through mechanotransduction.

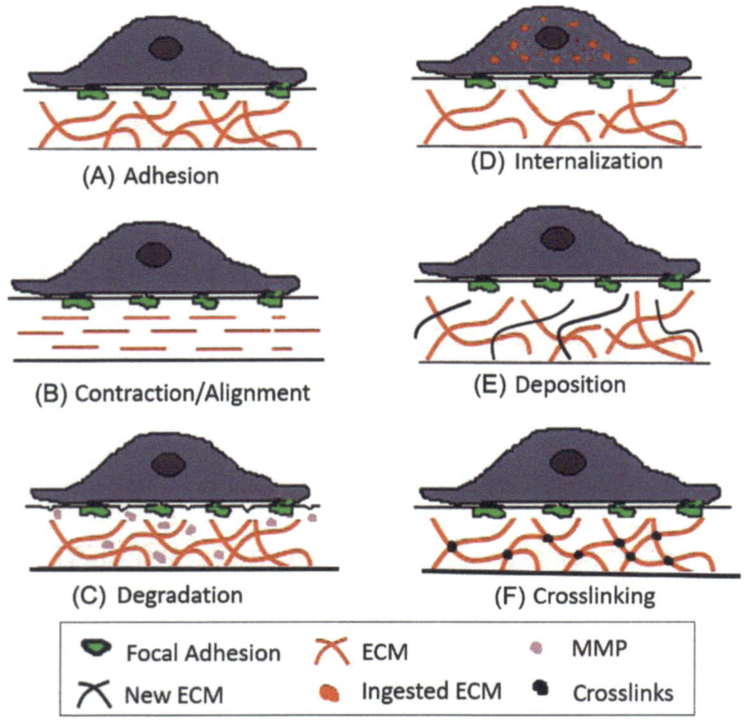

Figure 4.8 Mechanisms of ECM remodeling. The ECM is constantly remodeled depending on cell function. Mechanisms implicated remodeling of the ECM critical to proper cell function are schematically illustrated as (A) adhesion, (B) contraction, (C) degradation, (D) internalization, (E) deposition, and (F) cross-linking. *ECM*, Extracellular matrix; *MMP*, matrix metalloproteinase.

As emphasized earlier, when any imbalance in the deposition and the proteolytic processes in the ECM is shifted for a longer duration without a quick return to homeostasis, a serious progression of disease takes place. In the case of a higher rate of deposition of collagen and other fibrous proteins combined with the inability of the ECM to rid itself of these through the action of proteases, an accumulation of the fibrous proteins because of this remodeling occurs resulting in pathogenesis. This causes the development of various fibrotic diseases such as cirrhosis, cystic fibrosis, and cardiovascular diseases, leading to organ dysfunction. Increased ECM synthesis, which leads to its accumulation, is an abnormal reaction to injury to a tissue. Normally, the initial stage of wound healing is characterized by the attraction of fibroblasts to the wound site. An excess of fibroblasts, their differentiation into myofibroblasts which then deposit ECM in excess leads to tissue fibrosis. The excessive amount of collagen in the ECM increases tissue stiffness, which progressively leads to organ dysfunction. For example, in normal ECM synthesis in the lung tissue, both collagen, which is stronger and stiffer, and elastin, which is more elastic and compliant are deposited in the tissue. If the collagen deposition is dominant, the tissue stiffness changes leading to pulmonary fibrosis. This stiffness change has been implicated in the progressive attraction of fibroblasts with a myofibroblast phenotype, which then deposit even more collagen in the tissue and cause disease progression. This remodeling of the ECM upregulates the presence of TGF-β known to cause arterial hypertension and cardiovascular diseases. Prolonged hypertension alters the mechanical properties of the cardiovascular tissue, including stiffness, which is the stimulus necessary for disease progression by setting the stage for local invasion of fibroblasts and an increase in the deposition of the ECM causing tissue scarring.

It is now quite well established that tumor tissue is stiffer than normal tissue. LOX, also known as lysine-6 oxidase, has been implicated in the alteration in the normal rates of ECM remodeling. LOX elevates the rate of cross-linking of newly synthesized collagen in the ECM, clearly resulting in increased stiffness of the tissue. This raises the ability of cancer cells to become invasive indirectly a result of the ability of integrins to cluster and activate FAK, which, in turn, promotes mechanotransduction and provides a way for cancer cells to migrate. Cancer cells appear to migrate on collagen fibers, which have been linearized because of increased stiffness facilitated by the malignant transformation and become invasive leading to metastasis. A malignant tumor is generally surrounded by dense, fibrous tissue because of augmented ECM deposition while such a stiffened tissue is not observed in benign tumors. The breakdown of the normal ECM during remodeling and the formation of this fibrous tissue drives tumor progression at the primary site. Although the concept of metastasis as a natural spread of cancer cells in the body is still accepted, new perspectives on ECM remodeling have been discovered. The critical role of ECM remodeling not only in the migration of cancer cells from the primary site but also in the creation of appropriate niches in the secondary sites for tumor cell colonization and growth is now clearly recognized.

Remodeling of the ECM does not lead only to disease progression. The changes observed in stiffness in different tissues is a normal aspect of development. Cell differentiation is also a result of the variations in biomechanical properties of the ECM in various tissues. Both motile and nonmotile cells sense the properties of the ECM by exerting contractile forces against this substrate, which either results in cell migration or simply appropriate adhesion. One must not forget that there is a direct connection between the nucleus and the ECM via the cytoskeleton, and any reorganization in the ECM mechanically affects the chromatin arrangement in the nucleus and rapidly affects gene transcription. Alterations in the ECM properties under pathological conditions have a direct effect on cell–ECM interaction and as a result affects cell behavior. For example, if the structural fibers in the ECM are aligned in a certain orientation, cell migration is enabled. However, if the protein fibers form a tangled matrix or a dense cross-linked network, the consequent arrangement is now designed to decrease or prevent cell motility.

It is critical to understand that the cell—ECM interaction is reciprocal. There exists a feedback loop, which regulates any changes in the composition of the ECM by the action of the cells, which, in turn, are highly adaptive to their environment in carrying out their physiological activities. Although the ECM has innumerable other functions, which can be described in greater detail, the emphasis here is to relate ECM activity only to biomechanical influence.

Review and Project

1. What are the types of stresses that are applied to cells in vivo? Choose the knee cartilage and an artery and analyze the different stresses which cells in these tissues experience.
2. Research various methods of promulgating MSC fate. How do these compare with the application of mechanical stimulus to reach the same objective?
3. Can you think about methods to use mechanical stimuli in cancer therapy? What are the possible hurdles?
4. Research the effect of microgravity on cell proliferation. What novel therapies can be provided to astronauts after their return from space?
5. How can the modeling and remodeling processes of the ECM be regulated using mechanical stimulation?

CHAPTER 5

Tissue mechanotransduction

5.1 Introduction

In the previous chapters, the focus has been mainly on the effect of mechanical stimulus on individual cells, although in vivo coordinated effort in mechanotransduction has to occur for the tissue and, more universally, for an organ to respond to the external stimulus. Both flight and fight reactions were given as examples of this response in an earlier chapter. Nevertheless, in the following chapter, the reactions of specific tissue types to the application of mechanical forces are discussed in some detail. There is a paucity of information in the literature regarding tissue mechanotransduction. However, there is no denying the fact that various tissues are subjected to mechanical stimuli. Of course, we do not think about mechanotransduction in terms of a specific tissue or organ but more at a cellular level. Ultimately, the physiological response, which is observable, is many times at the tissue or organ level and becomes relevant in clinical medicine involving the treatment of disease.

The role of molecular cues, which are mechanical in nature in tissue and organ development, is widely accepted in organogenesis. In contrast, the importance of mechanical stimuli in initiating and controlling critical cellular processes in organ development has only recently been brought to light. In such a concept, rather than only individual cells, groups of cells of a single type or, even more commonly, groups of cells with various individual functions operate as networks to bring about organogenesis. Tight coordination between cells is imperative to their successful function in organs and in achieving morphogenesis. Hence, responses from such networks composing a tissue or an organ become critical despite the fact that the interplay between mechanotransduction and reaction may be complex. Clearly, there are a number of signaling pathways by which mechanical forces are transmitted between cells in an intercellular manner. For tissues or organs, one can talk about sensor circuits in lieu of individual cell receptors. In doing so, it is still unclear if a particular group of cells jointly acts as a first receptor or if the stimulus is itself experienced by individual cells, which then formulate and provide a collective response. Mechanical forces on cells and consequently tissues may be (1) intracellular as in the transmission of contractile forces through the cytoskeleton; (2) intercellular, experienced by endothelial cells (ECs) both in the skin and in the vasculature; and (3) between tissues as in the case of breathing where expanding lungs push out on the ribs. Fig. 5.1 is a clear illustration of examples of such forces, which act on a group of cells or tissue during cell

Mechanotransduction
DOI: https://doi.org/10.1016/B978-0-12-817882-9.00001-X

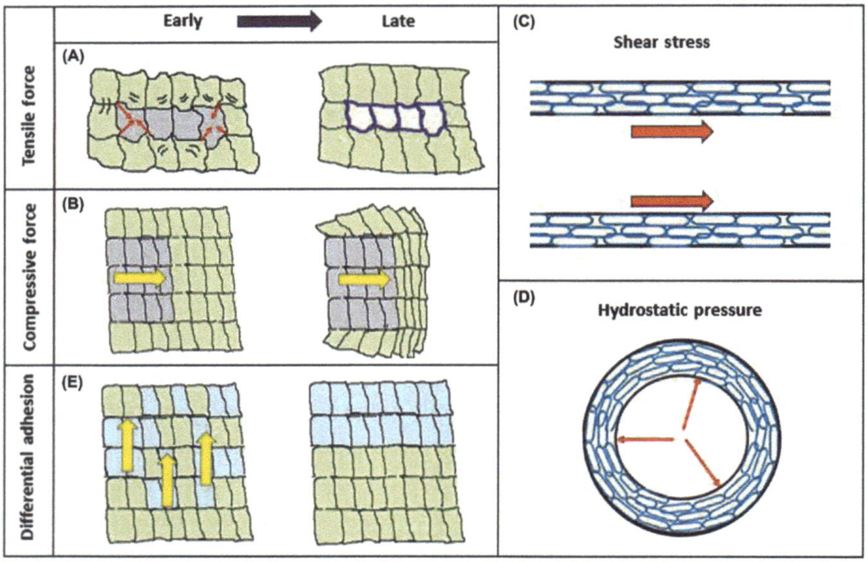

Figure 5.1 Types of forces acting on a group of cells or tissue. Forces acting on cells or tissue in vivo: (A) tensile, (B) compressive, (C) shear, (D) hydrostatic pressure, and (E) differential adhesion.

sorting, cell migration, differential adhesion, and forces exerted on adjacent tissue by the action of mechanical force on a certain tissue.

Tensile forces as shown in Fig. 5.1 are experienced by cells during dorsal closure, whereas neighboring cells are subjected to compressive forces during the growth of adjacent tissue. Shear stress due to fluid flow occurs on the surface of epithelial or ECs in vasculature and pulmonary airways. Hydrostatic pressure is experienced by smooth muscle cells (SMCs) during blood flow in the arteries, whereas differential adhesion and resulting tension occur during gastrulation. One can define these groups of cells that receive the mechanical signal as sensor circuits. Again, as in individual cells, this signal is transduced, resulting in effector functions that are not confined to individual cells. Rather, groups of cells perform the function of the effector in remodeling the extracellular matrix (ECM), bring about tissue differentiation and growth, and even cause large scale movement of the tissue or organ. A well-coordinated synchronization of the reorganization of tissue is extremely important in organogenesis during development. Tension forces are known to play a key role in tissue orientation in providing planar cell polarity and hence control cell division and changes in the shape of cells. Other processes of morphogenesis include branching, which are strongly influenced by mechanical tension. Fig. 5.2 illustrates this global sensor—effector circuit without mentioning the signaling pathways involved in the process.

The formation and growth of many calcaneal spurs appear to be a result of mechanical forces applied inadvertently or incorrectly to different tissues. Improper

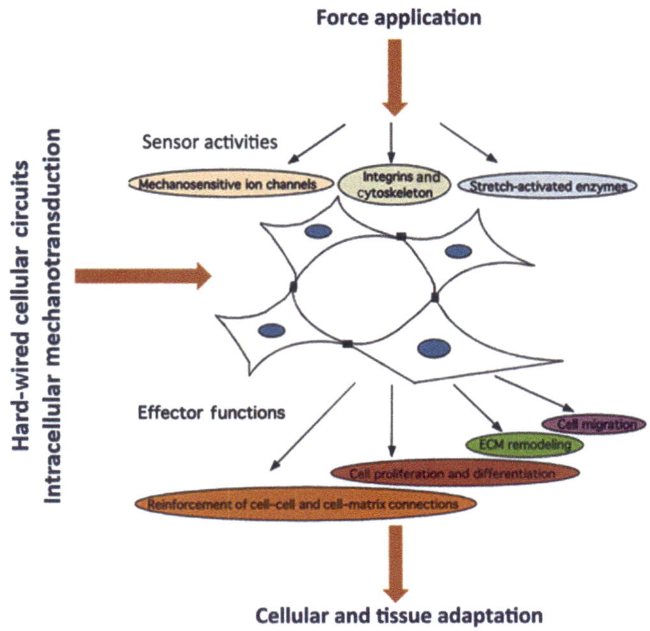

Figure 5.2 Sensor—effector circuits for tissue mechanotransduction. Tissue adaptation occurs at the macroscale, but this gross change is mediated at the cellular scale. Schematic illustration of the manner in which sensor—effector circuits effect tissue mechanotransduction and bring about tissue remodeling, although at the fundamental level, transformation is accomplished by the cells. Sensor—effector circuits involve the mechanosensory machinery, signal transduction pathways, and response from a group of cells or tissue. *Adapted from Ko, McCulloch. Intercellular mechanotransduction: cellular circuits that coordinate tissue responses to mechanical loading. Biochem Biophys Res Commun 2001;285: 1077–83.*

and excessive weight distribution from normal physiological formation or the incorrect use of footwear appears to cause the growth of these bone spurs in the human anatomy. Bunions may appear at the base of the big toe due to the accumulation of bone tissue and can be attributed to uneven weight distribution on the foot. Bone spurs also grow in the knee, in the spine, and in the iliac crest. Uneven or excessive load distribution appears to be the cause, although osteoarthritis has been singled out as the normal cause of these bone spurs. These mechanical force abnormalities are ubiquitous in our daily physical activities.

5.2 Signal characteristics

The mechanical forces experienced by cells are unique in that each signal is characterized by its magnitude, duration, and frequency. Contractile forces in cells appear to be pulsatory even during tissue morphogenesis, probably causing small incremental changes in the cell shape with each pulse. Thus it is still unclear if repetitive or multiple force signals are critically needed to propagate a stimulus across the tissue or organ or if just a one-time perturbance is sufficient to accomplish the required tissue response until the signal transmission is complete. It should be kept in mind that the degree of influence of a group of cells under the action of a direct mechanical force may be only as far as the neighboring cell or as much as 20–30 cells. Also, whether mechanical signals alone are more effective than biochemical signals and the extent of their outreach in terms of distance from the point of application to the response site is still a moot point. There is still much speculation about the actual mechanisms involved in mechanotransduction processes of this nature. Each of these signals is also uniquely recognized by specific cell mechanosensors. It is interesting to note that the forces generated by cells or groups of cells are regulated by a variety of mechanisms, which, in turn, are uniquely sensed by specific cell machinery to bring about the required physiological result. Mechanical loads transmitted through the intercellular junctions may also be important in embryogenesis and tissue homeostasis as observed in the formation of physiological defects and malformation and malfunction. The fact that cellular networks function as whole units is inevitably linked to their intercellular connections such as gap junctions and adherens junctions. Literature reports a number of observations from developmental defects in knock-out animals that lack these intercellular junctions linking them to embryo death resulting from malformations due to insufficiency in N-cadherin. Suffice it to say, intercellular adhesion complexes have the capacity to relay the force signal directly to neighboring cells sequentially and can bring about changes in the cytoskeletal machinery of these cells. This is accomplished through a rearrangement of actin filaments and intermediate filaments either in changing cell morphology or in coordinated movement of the tissue. While the signal to neighboring cells is enabled through the cell–cell communication machinery, force transmission through the adherens junctions may also directly apply loads on the nuclear membranes of neighbor cells and produce changes in protein and other transcription structures attached to the cytoskeleton to cause appropriate gene expression. Some examples of such well-known sensor–effector circuits include the coordinated action of osteocytes in acting as sensor cells to signal osteoblasts to mediate remodeling/modeling processes in the reabsorption of ECM debris or production of new osteoid. Fibroblasts also help in the remodeling of tissue because of mechanotransduction through the formation of cellular circuits that are connected via intercellular junctions. Movements over larger cell layers result in forces on adjacent tissue, which can be tensile or compressive in nature and can also affect ECM modeling. Differential cell proliferation may be responsible for the eventual shape of organs during organogenesis. Paracrine communication may also result in a group of cells forming a particular tissue from mechanical stimuli, although the signal transduction process itself may be biochemical instead of biomechanical.

5.3 Signal control: Structural and secretory factors

Two important parameters in the proper functioning of multicellular networks in tissue are temporal and spatial control. The precise timing involved in the appropriate regulation of the signal that is received by the cells in a tissue is key to the complete signaling process. A signal cannot be acting forever in such a manner as to receive a continued response beyond the temporal necessity of the tissue nor can it be so ephemeral that the correct response is not adequately obtained. For example, gap junctions transmit waves of calcium ions following the application of a force from one cell to its neighbors. Currently, there is no knowledge of a correlation between the flux of calcium ions and the amount of load applied. However, it is well known that these Ca^{2+} waves may be significant in the physiological rhythmic beating of the cilia in the lung. Also, changes in gene expression in the intermediate duration signals as well as cell proliferation in the long term have been associated with mechanical forces.

Spatial control is equally important in cellular networks and may involve structural and secretory factors. Paracrine secretions probably control the general tissue pattern, but adherens junctions as structural factors do play a key role in microscale level spatial control. The following structural and secretory factors influence temporal and spatial control.

5.3.1 Adherens junctions

Cell to cell connections and communications are accomplished by adherens junctions. These structures are principally composed of chemical bonds between transmembranal proteins of the cadherin superfamily, which form extracellular bonds between two adjacent cells. For example, E-cadherins are connected through α and β catenins in the intracellular domain to the actin cytoskeleton. Fig. 5.3 illustrates a typical adherens junction. These cell adhesion molecules make individual connections but appear to cluster together as the adhesion points mature and become very important in the transmission of mechanical signals from cell to cell. Thus these connections are dynamic in nature and critical to the formation, reorganization, and dispersal of cell—cell adhesions. In tissue development and transition from epithelial to mesenchymal transformation, adherens junctions play a key role as also in carcinogenesis where loss of cell—cell adhesion occurs.

5.3.2 Gap junctions

These are small intercellular canals that allow the transport of small molecules and ions between cells. Structurally, gap junctions are made up of proteins called connexins, which agglomerate into hexameral constructs called connexons. These gated conduits, which form between adjacent cells, are also activated by mechanical stimuli and are dynamically regulated. Some core functions include voltage regulation for neurons

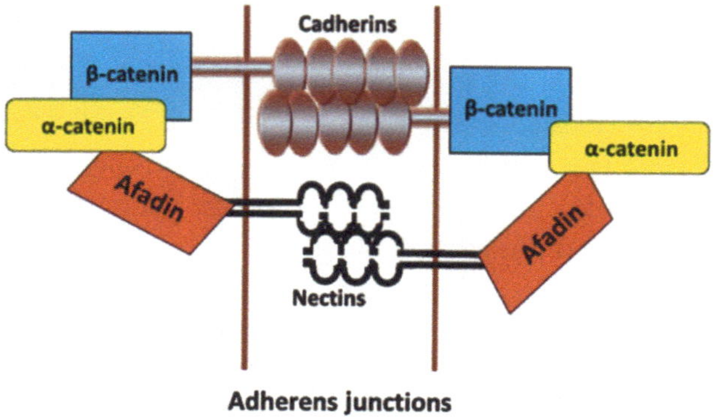

Figure 5.3 Structure of adherens junction. Detailed illustration of the structure of an adherens junction highlighting both transmembranal and intracellular bonding, which play a critical role in cell–cell communication of mechanical stimuli. Molecular machinery shows extracellular E-cadherins and intracellular β-catenins. These are involved in the transmission of forces between cells. *From Campbell HK, Maiers JL, DeMali KA. Interplay between tight junctions & adherens junctions. Exp Cell Res 2017;358: 39–44.*

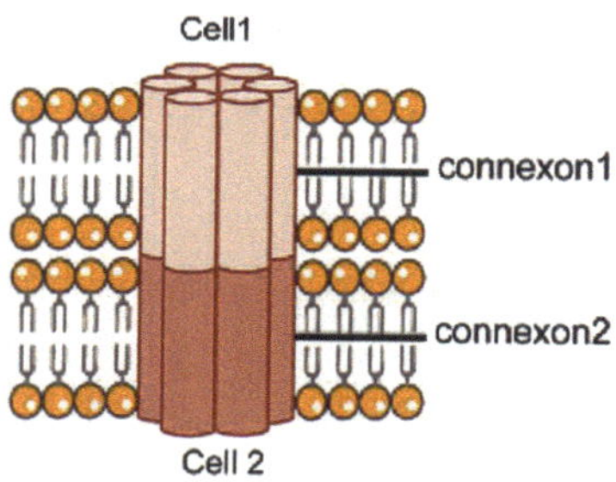

Figure 5.4 Structure of gap junctions. Molecular machinery in the cell–cell interface illustrating gap junctions, which are essential for the transfer of small molecules and ions through transmembranal proteins called connexons (green), which straddle the cell membranes of adjacent cells. *From Talbot J, Audrey Lamora A, Stresing V, Franck Verrecchia F. Gap junction in bone remodeling and in primary bone tumors: osteosarcoma and Ewing sarcoma. Bone Cancer (Second edition), Prim Bone Cancers Bone Metastases 2015: 83–89.*

and cardiac cells as well as a key role in vasodilation. Gap junctions serve to mediate metabolic demands in cells by allowing the transport of nutrients to maintain cell homeostasis and the movement of signaling molecules across groups of cells. The structure of gap junctions is illustrated in Fig. 5.4.

5.3.3 Secretory factors

Autocrine and paracrine factors were mentioned in an earlier chapter as modes of cell signaling. Such soluble, secretory factors mediate effective cell–cell signaling and stimulate as well as aid in providing a unified and coordinated response from a tissue.

Growth factors are also responsible for intercellular communication, which promotes tissue functions. For example, the augmented production of endogenous growth factors as a result of mechanical stimulation has a favorable effect on the development of fetal lung cells. The release of prostaglandin, prostaglandin E2, and interleukin-1, as a result of tension in the ligament, is known to result in periodontal bone resorption. Mechanical stimulus also causes an effective increase in the production of tissue growth factor in bone cells.

A group of cells or a tissue as such does not respond directly to external signals. Rather, it appears that the coordinated result of the individual cell receptors produces the overall response to the signal applied to the group of cells or tissue. Tissue mechanotransduction is extremely important in embryogenesis. Only recently are the mechanisms responsible for signal reception, transduction, and response being clearly identified in embryogenesis. One such mechanism is the induction of a signaling cascade by muscle contractions involving Rac GTPase and other proteins, including GITI (G protein—coupled receptor kinase, interacting ArfGAP I), β-PIX (also known as ARHGEF7), and p21-activated kinase. Other examples are hemidesmosomes, which are strengthened by the phosphorylation of intermediate filaments to protect epidermal cells against tension. Plectin, a protein that acts as a link between the cytoskeletal elements, can bind to the dystroglycan complex and relay mechanical signals in alveolar epithelial cells and can bring about morphogenesis and differentiation in these cells. Tension forces also restrict organ growth through a feedback mechanism through the Hippo mechanosensitive signaling pathway by regulating YES-associated protein (YAP)-I and TAZ (transcriptional coactivator with PDZ-binding motif) transcriptional factors.

In the following sections, we will focus on individual tissues and their interaction and response to mechanical stimuli, although influence at the cellular level is also included.

5.4 Musculoskeletal system

5.4.1 Mechanophysiological environment

The musculoskeletal system consists of hard tissue, made of many bone types and softer connective tissue, including muscles, cartilage, tendons, and ligaments. Other than its mechanical function of providing rigidity, integral protection, limb movement, and locomotion, bone is responsible for the formation of blood cells and storing calcium. At the cellular level, bone consists of two components: the organic portion comprising of predominantly collagen but also including elastin, proteoglycans, and other proteins; and an inorganic part, which includes nanoscale mineral particles of hydroxyapatite. Bone is critical in providing a support system for softer muscle tissue, the absence of which will result in a lack of shape and inability for any movement of the body.

This musculoskeletal system coordinates the transfer of forces, which, if considered carefully, is accomplished in many ways through mechanotransduction at the cellular level and precisely coordinated to bring about highly complex physiological functions.

Bone cells may be classified into three types: osteoblasts, which generate new tissue; osteoclasts, which digest older tissue debris; and osteocytes, which reside in bone tissue and nurture it in the long term. Osteoblasts mature into osteocytes and occupy lacuna in the bone tissues and produce and deposit a bone matrix material called osteoid. They communicate with each other through microcanals called canaliculi. Hematopoiesis is a significant and critical physiological function that is ascribed to this hard tissue for the production of various types of blood cells, including erythrocytes and leukocytes. The strength and toughness of the tissue are the results of a balance in the amount of calcium produced or eliminated to maintain mineral homeostasis.

Bone, based on its microscale physiology, can be differentiated into primary and secondary bone types. Primary bone requires a template on which it can be deposited and can be one of three categories: lamellar, plexiform, and primary osteon. Secondary bone, on the other hand, is the result of remodeling processes of mature bone or bone debris. Woven bone is sometimes included in this category, where the collagen fibers are not aligned or oriented in any particular direction. This type of tissue can be deposited without the need for a substrate and usually is a precursor to the formation of lamellar bone. Callus formation and growth of bone spurs are typical examples of the formation of woven bone tissue.

The structural characteristics of bone tissue innately influence its physical properties. The influence of mechanical forces on bone structure was enunciated well over a century ago. Even in embryonic development, the growth of bone tissue is reported to be strongly associated with forces that the embryo is subjected to during gestation. Growth and stability of bone are dynamic in nature, and epigenetic factors linked to applied loads through exercise play a key role in mineral homeostasis and the resulting integrity of bone structure. The complex bone architecture and its functional adaptation evolve based on the mechanical loading patterns on this tissue, even in adults. This is thoughtfully reflected, albeit conjured in a less physiological sense, in Wolff's law of functional adaptation, which paraphrased as, "static stressing determines the shape of bone *macroscopically* (emphasis added by author) and only static function and necessity determine the location of every bony element and overall shape of bone *microscopically* (emphasis added by author)." Despite the strong role of hereditary factors in the evolution and growth of bone tissue in an individual, there is no gainsaying the fact that over a number of years, there is irrefutable physical evidence that mechanical loads have a singular influence on the shape and function of bone even at the cellular level. Therefore the fact that the lack of sufficient and appropriate mechanical loading on bone tissue can give rise to pathology and disease, including osteopenia and more severe osteoporosis, cannot be overstated.

From a mechanics point of view, bone is highly anisotropic, and an attempt in a load or stress analysis using trivial static stress analysis will result in incorrect conclusions. However, for an inclusion of microscopic features in load analysis, it is imperative that a multiscale computational model is more appropriate. This book, in no way, provides a complete mechanical treatment of bone tissue at the macroscale. Rather, the focus is on the effect of loads at the cellular and tissue level. Fig. 5.5 is an illustration of the structure of bone.

Two types of bone formation mechanisms are available: intramembranous ossification and endochondral ossification. In the former, bone tissues such as those in the skull and other flat bones are generated directly through the anabolic process by the differentiation of mesenchymal stem cells (MSCs) to osteoblasts. In the latter mechanism, MSCs are differentiated into chondrocytes, which form soft cartilaginous tissue, which is further mineralized by osteoblasts following chondrocyte apoptosis. Such processes occur in long bones.

Muscle, on the other hand, is a connective tissue, which actually transmits loads and forces by contraction to accomplish various physiological functions. In one sense, the reverse process corresponding to mechanotransduction takes place in muscle at a gross scale, since in this case, chemical energy is converted to mechanical movement. Muscles are hierarchical structures composed of bundles of fibers, which vary in length and diameter. Each muscle is surrounded by a fascia and furthermore by a sheath called the epimysium. Each muscle in itself is made up of bundles of myofibrils or fascicles covered by a connective sheath called the perimysium. A single myofibril, in turn, is

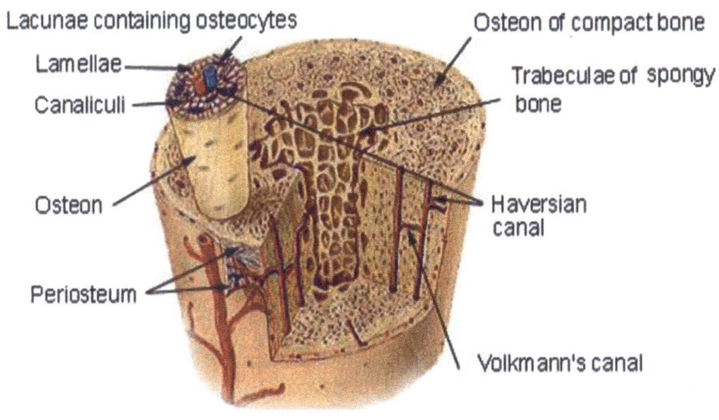

Figure 5.5 Structure of bone. Structure of bone tissue illustrating compact (outside circumferential region) and spongy (inner core region) bone. Osteon units formed by concentric lamellae with individual Haversian canals can be observed with the system of blood vessels. Individual osteocytes in lacuna are connected to neighboring osteocytes through canaliculi. *From SEER Training Modules. Structure of Bone Tissue. U. S. National Institutes of Health, National Cancer Institute. <https://training.seer.cancer.gov/>.*

composed of sarcomeres, each of which can be considered as the basic unit of the muscle. Both thick filaments made up of myosin and thin filaments composed of actin together form a sarcomere. Bundles of sarcomeres acting in coordination cause muscle contraction, which is the fundamental cause of the movement of the skeleton. Fig. 5.6 is an illustration of the force–length curves corresponding to muscle contraction and relaxation for skeletal and cardiac muscles.

There are generally two types of muscles, namely, striated and nonstriated or smooth muscles. The latter are controlled by the autonomic nervous system, while the former are subjected to action by direct voluntary movement. The emphasis in this section will be only on the striated muscle since it is directly linked with force application and hence the resulting effects of this mechanical stimulus as such. No attempt is made to study the other soft tissues, such as ligament, tendon, or articular cartilage. Two properties of muscles, which are quite important in understanding function as well as calculating loads on the musculoskeletal system, are the force–length and force–velocity relationships. These properties differ between the various muscle groups based on the structure and functional demands of the physiological activities specific to each one of them.

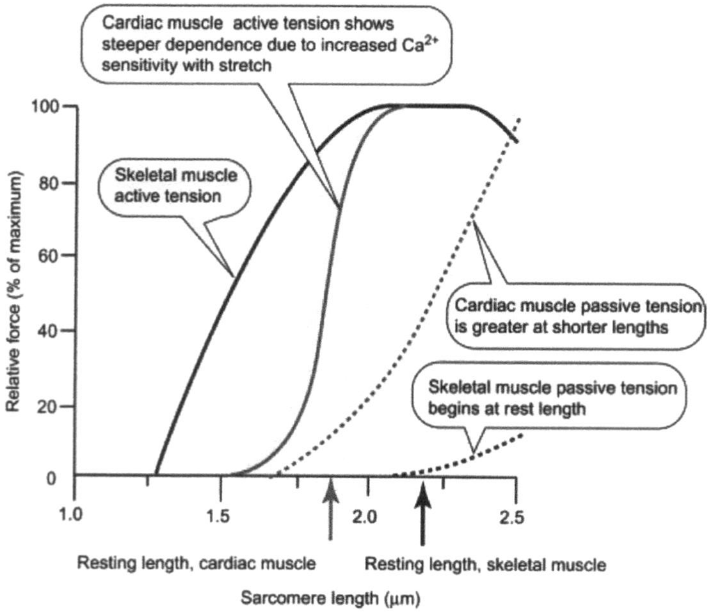

Figure 5.6 Force–length curves for muscle. Typical force–length curves for muscle extension and contraction for skeletal and cardiac muscles. The relative force magnitude (in percentage) as a function of sarcomere length can be seen for the two muscle types. The force generated by the muscle fiber is clearly zero at its resting length and reaches a maximum when muscle contraction occurs. *From Feher J. The cellular basis of cardiac contractility. In: Quantitative human physiology, 2nd ed.; 2012, 547–55.*

Force—length property refers to the maximum force that a muscle can exert with regard to its length. This, in turn, translates to the number of cross-bridges that exist along the length of each sarcomere. Since each cross-bridge generates the same amount of force and work, and since the cross-bridges are arranged equidistant along the length of the sarcomere, the larger the number of cross-bridges along a fixed length of the sarcomere, the higher is the force generated by this fiber length. Thus shorter fibers with a relatively large number of cross-bridges are more prone to generate a higher force. Again, there are limits to the shortness of the length of the fiber since overlapping of cross-bridges annuls their ability to generate force. Fig. 5.6 shows hypothetical curves for derived force—length curves for skeletal and cardiac muscles. Force—length property of muscles is also subjected to adaptation based on the functional demands of the specific muscle. For example, the rectus femoris muscles of bicyclists are stronger at short muscle lengths compared to runners who have stronger muscles at longer lengths, highlighting adaptation to positional use during exercise.

Force—velocity property is attributed to the relation between the maximum force generated by a muscle to its instantaneous rate of change in length (velocity). These properties are determined for maximum activation at optimal sarcomere length. The force—velocity relationship for a skeletal muscle is shown in Fig. 5.7, where it can be seen that peak power is attained when the speed of muscle shortening is about 30%—35% of the maximal unloaded shortening speed.

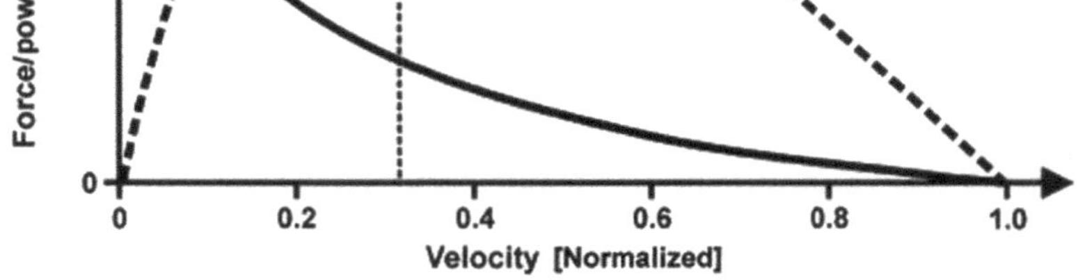

Figure 5.7 Force—velocity relationship for muscle. Typical force—velocity relationship for a skeletal muscle originally attributed to Hill (1950). The corresponding power curve is also shown. Note that peak power in a muscle is produced for the optimum force corresponding to 30%—35% of the maximum muscle shortening velocity. *From Herzog W. The biomechanics of muscle contraction: optimizing sport performance. Sports Orthop Traumatol 2009;25:286—93.*

5.4.2 Mechanotransduction in musculoskeletal tissue

5.4.2.1 Bone tissue

It is well known that individual cell shape is altered by mechanical stimulus on its environment. This, in turn, regulates cell expression, which affects the biochemical environment around the cell and therefore around the surrounding tissue. Such an interactive loop ultimately results in some sort of equilibrium between the tissue and its mechanical function. Four key steps have been identified during the regeneration of bone tissue: (1) grouping together of MSCs, (2) their differentiation and subsequent collagen expression, (3) calcification of the tissue and corresponding angiogenesis, and (4) proliferation of osteoblasts and formation of bone tissue. Functional adaptation of bone tissue to mechanical stimuli normally includes bone growth or changes in the bone structure. Both surgically invasive and noninvasive studies using knock-out or knock-down animal models have provided sound proof of the critical necessity of mechanical stimuli for bone growth. Such a response to mechanical loads by cortical bone has resulted in some well-established paradigms linking mechanical loading to cortical bone formation.

1. Formation of cortical tissue occurs as a mechanoresponse to dynamic loading but not to static loading.
2. For cortical bone formation to occur, a minimum threshold strain must be applied. As a corollary, mechanical loads, which produce corresponding strains below this minimum threshold, are ineffective in causing the formation of cortical bone.
3. Cortical bone formation reaches a plateau after a relatively low number of loading cycles.

In contrast, it appears that the minimum force threshold is essential for the formation of trabecular bone, although reports on the sufficiency of static loading for trabecular bone growth have at best been conflicting. In retrospect, the validity of the data obtained from surgically invasive testing methods is always questionable in light of its effect in modifying or altering scientific observations.

Considering that dynamic loading is extremely important for cortical bone formation and growth, the magnitude, frequency, and the continuity of application of this mechanical stimulus on the tissues become physiologically relevant. First, magnitude refers to the highest load or strain applied during a specific stimulus and is strongly correlated to bone formation. Nonetheless, the same peak stimulus may not be relevant to all cases and will vary depending on species and within the species, age, sex, weight, genotype, etc. In general, a 1200–2000 $\mu\varepsilon$ range appears to lie within this peak strain magnitude across a wide variety of species. It should also be kept in mind that during bone modeling and remodeling, the strain values will change, and hence, the functional adaptation of bone tissue may vary from case to case. The second factor, frequency, appears to have a corresponding threshold below which there is a dearth of bone tissue formation. At frequencies of mechanical stimulus of less than 1 Hz, the effect on bone tissue is akin to the static loading case. On the other hand, a high frequency of 10 Hz results in a deleterious effect on tissue generation or no effect at all. Optimum stimulus corresponds to a frequency range of 2–5 Hz, which falls within the gamut of physiological values. It is also reported that rest between periods of dynamic loading is essential for the growth of bone tissue. The duration of the rest period has not been quantified, except that cells become desensitized if loaded continually.

Various types of mechanical stimuli have been reported to cause bone tissue formation. These include strain, pressure or compressive loads, and shear stresses resulting from fluid velocity. Fig. 5.8 is an illustrative hypothetical concept map indicating mechanical parametric considerations as a function of time, under which bone tissue formation is favored in vitro. This map provides hypothetical conditions where preosteoblasts may differentiate faster above certain thresholds. These maps may provide guidance to bioengineers and biologists regarding the conditions required to generate bone tissue from undifferentiated cells or even stem cells. This concept map may be extended to include other cell types and other stimuli or experimental conditions, including stiffness, frequency of applied stimulus while using mechanobiological regulations, that summarizes the formation of soft and hard tissues based on the different combinations of these mechanical stimuli. Briefly, under conditions of low strain and

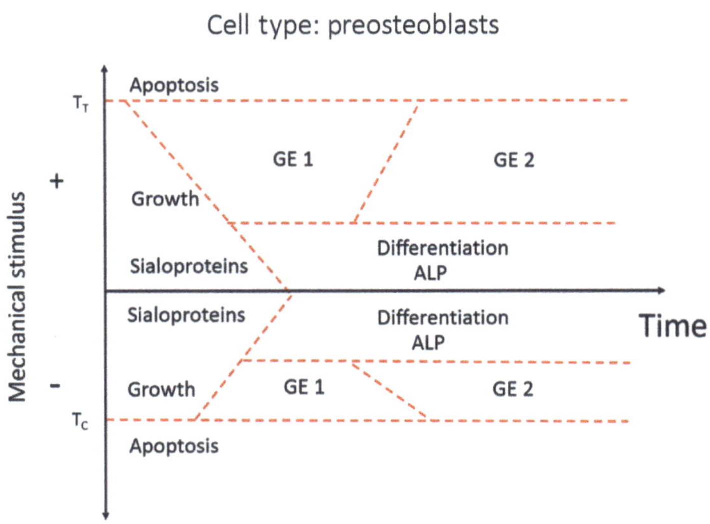

Figure 5.8 Concept map of cell response. Concept map of possible cell responses for a specific mechanical stimulus as a function of time. *ALP*, Alkaline phosphatase; *GE*, gene expression; T_T and T_C, threshold values of the stimulus in tension and compression, respectively, above which apoptosis occurs. This is hypothetical for preosteoblasts and not based on experimental evidence.

high compression, cartilage formation is preferred, whereas, in the high-strain regions, connective tissue and fibrocartilage are formed. For distraction osteogenesis where the bone tissue is fractured on purpose and placed under a tensile load, dilation rather than compression is the preferred growth environment. It must be noted that under normal fracture of bone, healing occurs effectively under a compressive load. Also, it is important to combine this knowledge with the three paradigms mentioned earlier to understand functional bone adaptation in light of the response to mechanical forces. These concept maps must reflect the changes that occur in the cells under a number of in vitro conditions.

5.4.2.2 Cartilaginous tissue

Similar to the four steps for bone tissue regeneration described in the previous section, the regeneration of cartilaginous tissue also follows this process except that instead of differentiation of MSCs eventually into osteocytes, in this case, MSCs convert to chondrocytes. The avascular nature of cartilage renders it difficult to regenerate in situ after damage or deterioration as a result of osteoarthritis. Fortunately, the potent nature of mechanical stimulus as a regenerative agent of this tissue provides opportunity to fabricate cartilage ex vivo. Based on ample empirical data, it is widely accepted that mechanical loads can stimulate the biosynthesis of cartilage ECM. However, the current understanding of the influence of mechanical stimulus on cartilage regeneration is fairly empirical. Besides, there is a lack of consensus on the type of stimulus or its magnitude, which will render cell function most effective in the process of regeneration of the tissue. The cell signaling mechanisms with regard to mechanotransduction are also not clearly delineated, an understanding of which may be critical to tissue regeneration ex vivo and the adoption of regenerative therapies.

One interesting side note that is highlighted here is the influence of mechanical stimulus on tissue formation even at the tissue—implant interface. Some type of tissue regeneration occurs at an implant interface due to mechanical stimulation as a result of micromotion at the implant interface. During the healing process, the MSCs that arrive at the implant—tissue interface are either differentiated into fibroblasts that produce an undesirable fibrous connective tissue or osteoblasts that deposit the bone tissue matrix to enhance osseointegration. Other modes of tissue generation probably occur at other tissue—implant interfaces resulting from mechanical stimulus.

5.5 Pulmonary system

5.5.1 Mechanophysiological environment

The pulmonary system consists of the respiratory pathway, including the trachea, the bronchi, and the lungs. This system is indeed dynamic and is subjected to constant changes during normal respiration but experiences more drastic changes during heightened physical activity. Thus the cells that comprise the many tissues that are part of the pulmonary system are constantly imposed upon by loads, which lead to stretching or extensional strain due to lung expansion. Besides, the epithelial surfaces of the respiratory pathways are also subjected to a shear stress because of the flow of air in contact with them. Depending on whether normal breathing or abnormal respiration as a result of increased physical activity is occurring, the cells that make up the tissues in the lung experience various magnitudes of loads and stretch. These include the lung parenchyma and the airways, although the pulmonary and bronchial vasculature must also be included in this system. It must also be noted that vascular perfusion occurs at the alveoli where exchange of gases takes place, which may involve a completely different type of mechanical stimulus.

The blood pressures in the pulmonary artery are much lower than systemic values with about 20/12 mm Hg during the systolic/diastolic phases. Correspondingly, the pulmonary arterial walls are much thinner and for good reason since they are separated by a thin barrier of about 0.5 μm thickness from the alveoli where the gas exchange occurs. The barrier itself consists of the epithelial cells from the alveolus, the basement membrane, and the arterial epithelial cell layer. The stresses in this layer are clearly consonant with its thickness, vis-a-vis the corresponding stresses in the thicker and multilayered systemic arterial walls. Fig. 5.9 is a concise diagram of the pulmonary system with the airways clearly delineated up to the alveoli.

The diaphragm and the intercostal muscles are responsible for the expansion and collapse of the lungs during respiration. During inspiration, as the diaphragm contracts, it pulls the inferior portion of the lung downward, allowing air to flood into the lungs.

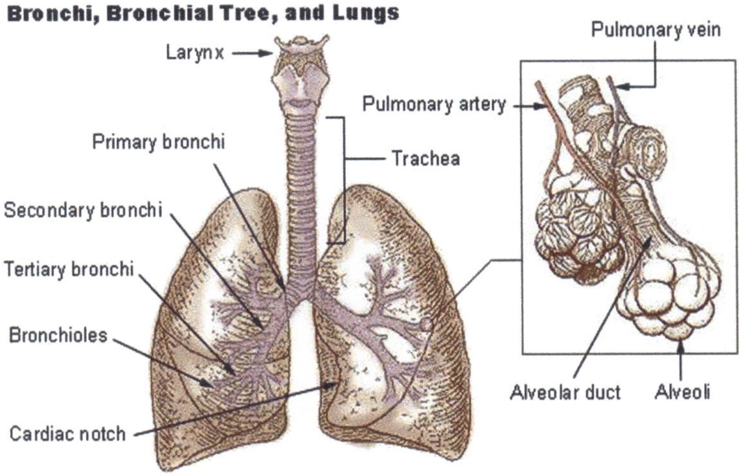

Figure 5.9 The pulmonary system. The lower respiratory tract including the trachea and bronchi, bronchioles, and alveoli in the lung. Inset shows a few details of the alveoli. Gas exchange takes place at the alveoli. *Retrieved from www.physiopedia.com*

Expiration is a result of the diaphragms and the intercostals muscles relaxing and returning to their normal positions resulting in an expulsion of the air in the lungs as they contract and return to their residual volume. The variations in the volume of the lung during normal breathing and strenuous exercise affects not only airflow but also metabolic processes that take place in this organ. During early development, the physical forces on the lung appear to affect its function. In well-developed pulmonary systems, a varied range of mechanical stimuli is applied to different regions. The alveolar epithelium experiences strain during breathing, while shear stress acts mainly on the vascular endothelium, although the latter may also be subjected to strain and hydrostatic pressure. Shear stress by way of the pleural fluid affects pleural mesothelial cells, whereas the airways and the alveolar epithelial cells are subjected to shear stress from air during the respiratory process. Various parts of the cell are subjected to different stresses depending on the application site. As an example, shear stress from the fluid layer acts on the apical surface of the epithelial cells in the airway and alveoli. On the other hand, distention of the basement membrane results in a stretching of the basolateral surface of the epithelium. It must be noted that all these stresses are a local effect of the overall complex respiratory process whose characteristics can be simply measured by lung capacity. Of course, lung capacity and efficiency vary, and the typical volume—pressure relationship for an isolated lung is shown in Fig. 5.10.

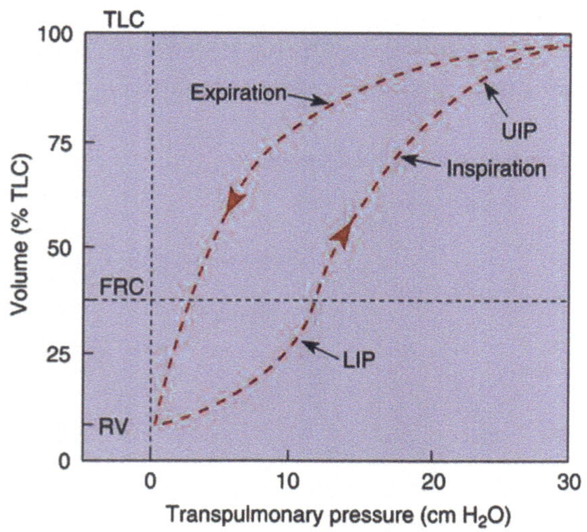

Figure 5.10 Volume—pressure relationship in the lung. Volume—pulmonary pressure curve for a lung showing the inspiration and expiration paths. UIP and LIP refer to the upper and lower inflection plots between which the tidal volume during mechanical ventilation should be limited. *FRC*, Functional residual capacity of the lung after a natural exhalation; *RV*, residual volume of air that is always present in the lungs; *TLC*, total lung capacity. *From Beachey W. Respiratory Care Anatomy and Physiology: Foundations for Clinical Practice. 4th ed.; 2018.*

5.5.2 Mechanotransduction in pulmonary tissue

As in the other systems, the mechanotransduction process involves mechanosensing, signal transduction, and response. This overall process includes the activation and complex interplay of a number of signaling cascades, secondary messengers, phosphorylation, signal amplification, and gene expression, which are regulated by transcriptional factors. Lung tissue stretching upregulates nuclear factor kappa B (NF-κB), which appears to possess a shear stress response element in its promoter region. Again, the possible mechanisms of mechanotransduction and possible responses from the cells in the complex pulmonary system may vary considerably.

Three mechanotransduction mechanisms are proposed to act in the pulmonary tissue. The mechanosensors in the pulmonary system include the stretch-activated ion channel, which regulates Ca^{2+} influx as observed in arterial SMCs. A second pathway is the plasma membrane stress disruption pathway, where excess stress can cause damage to the integrity of the plasma membrane. This leads to changes in Ca^{2+} homeostasis and affects signaling pathways causing protein kinase C activation and also upregulation in the expression of *c-fos*, resulting in binding of the stretch response element in early response genes to activate transcription. A third mechanism is the matrix—integrin—cytoskeleton pathway. In a general sense, integrins sense the mechanical stimulus through interaction with the ECM. They then relay the signal via the focal adhesion complex, the actin filaments, which serve as stress fibers and on to the nucleus. Eventually, NF-κB is activated for subsequent translocation to the nucleus to initiate the gene transcription process. The focal adhesion complexes are closely associated with a variety of molecules, which are responsible for the transduction, including connector proteins, tyrosine kinases, oncogene regulators, and a few growth factors receptors. There is still much to be learned about mechanotransduction in the pulmonary tissue and knowledge about the signaling cascades during this process is still not clearly grasped yet.

As mentioned earlier, cell—cell communication is critical to the transmission of the mechanical stimulus. This may involve a variety of communication modes discussed in earlier chapters and includes cytokines, growth factors, and soluble factors. These may play a role in enhancing the function of fetal lung fibroblasts, stimulate the secretion of a differentiation factor, and also increase the production of surfactant proteins in alveolar epithelium. In fact, applied strain as a result of fetal breathing is the factor indicated to cause the growth of fetal lung cells. Mechanical forces also are instrumental in the release of inflammatory mediators in lung cells. As in other tissues, the communication channels between the ECM and the cell modulate the mechanical signals sensed in the ECM and transmit them to the cell interior. Also, intercellular junctions such as tight junctions and gap junctions connect neighboring epithelial cells in the alveoli into a sheet, which helps transmit force by means of the intermediate filaments.

Changes in the ECM are caused in most cells in lung tissue due to mechanical forces including gene expression and protein synthesis of collagen, glycosaminoglycans (GAGs), and proteoglycans (PTGs). However, this protein distribution may vary depending on the oscillatory pattern of application of the mechanical stimulus, emphasizing the importance of these temporal factors.

5.6 Cardiovascular system

5.6.1 Mechanophysiological environment

The cardiovascular system comprises of the heart and a hierarchical network of blood vessels, each with a distinct structure. The arteries carry blood from the heart to the arterioles, which, in turn, feed capillaries where exchange of oxygen, carbon dioxide, nutrients, and waste occurs at the tissue level. This blood is then collected by venules and returned to the heart through veins. In general, the blood vessels are composed of various layers of cells, which form the vascular lumen. This structure is lined with ECs in the innermost layer called the tunica intima, which is in contact with blood. The next layer is the tunica media, which is made up of SMCs, which in the larger vessels consists of more than one circumferentially oriented layer. The outermost layer, the tunica adventitia, consists of ECM, collagen, fibroblasts, and progenitor cells and is supported by an external elastic lamina. The larger the blood vessel, the greater the number of circumferential SMC layers, while in smaller capillaries, these are replaced by a discontinuous layer of pericytes whose structure facilitates the exchange of nutrients and waste across a thin membrane. The vasculature in each tissue and organ is structure specific to meet the needs of each tissue and is constructed in such a manner through morphogenesis. A monolayer of ECs lines the entire vasculature and is in intimate contact with the flow of blood. The complex composition of blood and its components by virtue of fluid flow over the EC layer subjects it to shear stress. SMCs make up each layer of the tunica media and control the flow of blood through contraction and relaxation, by means of cytoskeletal proteins. This dynamic nature of the mechanical stimulus in the vasculature has been associated with pathologies as a result of age or extreme physical activity. The vascular ECM is composed of collagen, elastin, microfibrils, glycoproteins, and proteoglycans and provides tensile strength and elasticity to the blood vessels. Fig. 5.11 illustrates the cardiovascular system, including its principal components.

The cardiovascular system is ultimately responsible for the circulation of nutrients to all tissues in the body and the recollection of waste from these tissues. The heart pumps blood to all these tissues through two distinct loops. One loop transports blood between the heart and the lungs, while the other carries blood from the heart to the other tissues in the body. This transport is achieved via the pressure gradients originating at heart and carried over to the arteries, arterioles, and capillaries in the

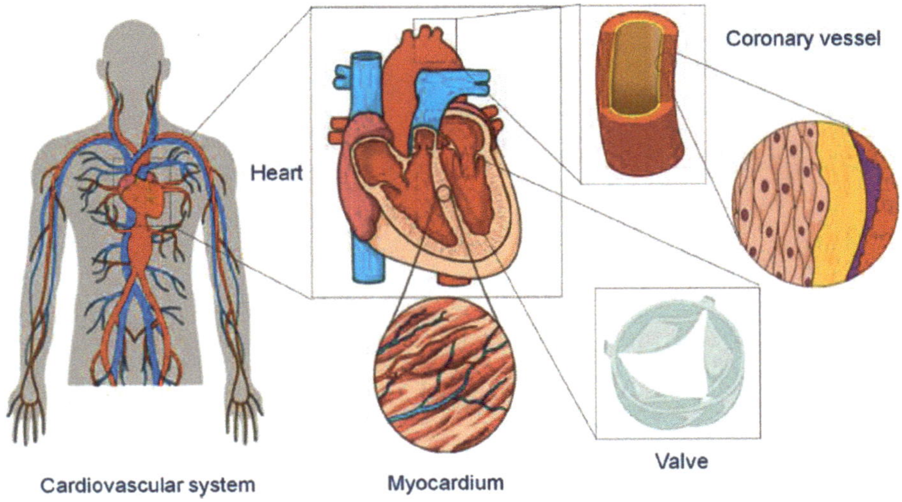

Figure 5.11 The cardiovascular system. The cardiovascular system consists mainly of the heart and the associated vasculature. The arteries carry oxygenated blood from the heart, which pumps through the functioning of the cardiac muscle. *From Cui H, Miao S, Esworthy T, Zhou X, Lee SJ, Liu C, et al. 3D bioprinting for cardiovascular regeneration and pharmacology. Adv Drug Deliv Rev 2018; 132:252–69.*

forward direction and through venules and veins as return pathway. The pumping of blood is accomplished during the cardiac cycle where the atria fill up with blood during the diastolic phase, which is then pumped out by constriction of the ventricles during the systolic phase. In the average adult, the heart beats at about 75 beats/min, with a stroke volume of 75 mL/beat. Under intense physical activity, the heart rate increases up to 150 beats/min and the stroke volume to 130 mL/min. In the aorta, the velocity of blood is about 33 cm/s, which then decreases to 0.3 mm/s in the blood capillaries. Such a drastic change in velocity combined with the oscillations as a result of physical activity imposes variations in shear stress, which should directly affect the epithelial cells that line the blood vessels. Also, under normal conditions, the blood pressure in these vessels oscillates between a systolic pressure of 120 mm Hg and a diastolic pressure of 90 mm Hg. This pressure decreases to about 35 mm Hg in the arterioles to about 10 mm Hg in the venules. These data are concisely gathered in Fig. 5.12.

It is also important to take into consideration the increase in the velocity of the pressure pulse, as it is transmitted through the arterial system and its influence on the different layers of the lumen as a function of compliance of these vascular tissues and the corresponding changes with disease and age. The pulsatile flow of blood through these conduits and the very nature of the fluid with particulate matter in the form of erythrocytes, leukocytes, and platelets render the physical/mechanical environment

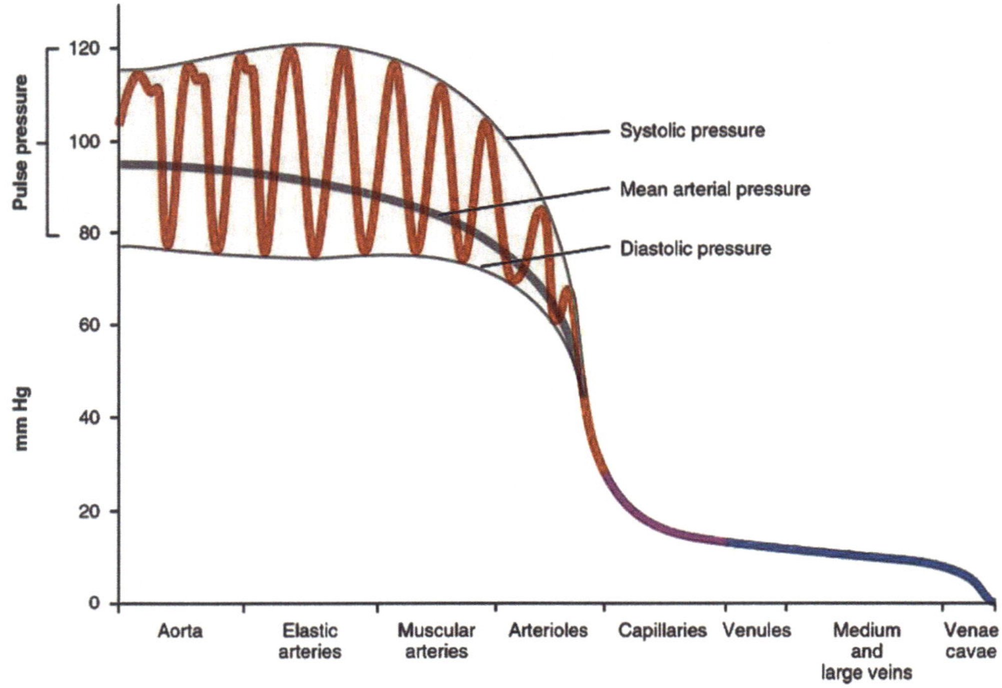

Figure 5.12 Blood pressure profiles in vasculature. Blood pressure in the different modules of the vasculature, indicating pulsatile pressure in the arteries. In contrast, venal pressure is relatively quite low. *From OpenStax College Anatomy & Physiology, Connexions Web site. <http://cnx.org/content/col11496/1.6/>; Jun 19, 2013.*

that the epithelial tissue interfaces with, quite complicated. The stresses and strains in the lumen not only vary across the cross-section of the blood vessel but also the variation of the stress magnitudes only adds to the severe mechanical stimulus experienced by these tissues. Another factor that must be taken into account in fluid flow is turbulence caused by the variations in fluid velocity as a result of intense physical activity. Bifurcations in the blood vessels may also be responsible for this flow phenomenon, resulting in a complex interaction with the epithelial tissue. Buildup of plaque will also cause local changes in fluid velocity, leading to localized cells of microturbulence. The phenomenon of mechanotransduction, therefore, becomes very complex when all these factors are taken together, especially when one considers the response of the individual cells that constitute the tissue or organ.

The cardiac muscle, including its walls, is subjected to a variational stress. Here, it is important to note that different regions of the heart may be experiencing different stress magnitudes at the same time. Thus instead of a uniform stress in the heart walls, the ventricle is undergoing systolic contraction, while the atrium is in the diastolic phase.

The blood pressures in the pulmonary artery are much lower than systemic values with about 20/12 mm Hg during the systolic/diastolic phases. Correspondingly, the pulmonary arterial walls are much thinner and for good reason since they are separated by a thin barrier of about 0.5 μm thickness from the alveoli where the gas exchange occurs. The barrier itself consists of the epithelial cells from the alveolus, the basement membrane, and the arterial epithelial cell layer. The stresses in this layer are clearly consonant with its thickness, vis-a-vis the corresponding stresses in the thicker and multilayered systemic arterial walls.

5.6.2 Mechanotransduction in cardiovascular tissue

A number of cells participate in the process of mechanotransduction in the cardiovascular system. These include ECs, fibroblasts, and cardiomyocytes. However, these do not share a common mechanism even if the applied mechanical stimulus is the same, and as stated earlier, the response to a specific stimulus can be totally different. When a mechanical stretch is applied to cardiac fibroblasts, angiotensin II activates mitogen-activated protein (MAP) kinases, including the $G\beta\gamma$ subunit of the G_i-protein and tyrosine kinases involving the Src family. In contrast, in cardiomyocytes, angiotensin II activates G_q and protein kinase C. In addition, different types of stretching have different effects in the same cell type. As an example, there is no c-fos gene expression in cardiomyocytes under hypotonic swelling in the presence of tyrosine kinase inhibitors. This effect is not seen for inhibitors of protein kinase C, phospholipase C, or angiotensin II antagonists. Conversely, c-fos gene expression is inhibited by tyrosine kinase inhibitors, protein kinase C, and phospholipase C under direct stretching.

The identity of molecular mechanisms of mechanotransduction in cardiovascular tissue is not well known although two types of models, centralized or localized and decentralized models, have been proposed. In the centralized model, three principal structures are choice candidates. First, stretch-activated channels have been suggested as good candidates for mechanosensors. When cardiac myocytes are stretched, the intracellular calcium levels increase lending credence to this model. Second, as in other cells, integrins are also good candidates as mechanosensors in cardiomyocytes and transmit the force signal via the cytoskeleton. Third, Src, which is a nonreceptor tyrosine kinase and a secondary messenger system, is also linked to sensing under mechanical stretch. In the decentralized models, the mechanical signal is applied remotely at the cell surface and is transmitted remotely through the cytoskeleton. The tensegrity model provides the basis for this decentralized mechanotransduction mode, which can occur at a point far away from the site of the applied signal. Increases in gene expression, protein synthesis, and cell size in myocytes support this model. Both localized and decentralized models may not be mutually exclusive. In fact, they may operate in parallel, and there may be cross-talk between the signaling pathways in these two models.

5.7 Nervous system

5.7.1 Mechanophysiological environment

As mentioned earlier, the ECM of the adult brain occupies 20% in volume of this organ. While a network of fibrous proteins pervades the peripheral soft tissues, the brain tissue itself is softer because its ECM consists of nonfibrillar collagen, which results in a relatively lower elastic modulus. The mechanical changes in the tissues control the production of ECM. Normal development of the brain requires the correct patterns of cell migration and appropriate differentiation schemes. This is completely dependent on alterations in the brain tissue ECM. In many tissues, fibroblasts and other mesenchymal cells are responsible for producing ECM, whereas in the brain, neuronal cells of all types, including neurons, glial and stem cells, contribute to ECM deposition. Organogenesis during embryonic development is also subject to mechanical stimulus. Gastrulation precedes organogenesis, and more specifically, neurulation starts with the migration of cells to the neural plate, which turns into the neural crest followed by the formation of the central nervous system (CNS) and peripheral nervous system (PNS). Migration of cells to the appropriate regions to accomplish organogenesis relies on mechanical signaling. All through the neurulation process, the conformal changes in the organ are controlled by mechanical forces. For example, stiffening of dorsal tissues is a result of contraction of cells, which enable closure of the neural tube, failing which a number of neural tube defects can develop. Also, during embryo development, the cells are cognizant in responding to the microenvironment in which they are present by means of mechanical signals and forces. Thus cell behavior and their organization within brain tissues are regulated through the mechanical signals that they receive and transmit during brain morphogenesis. Also, specific cell types develop and differentiate during embryogenesis based on the mechanical properties of the region in which they are located. Even for adult stem cells, the stiffness of their niche or their microenvironment becomes a critical factor in determining their fate and lineage. This niche stiffness is dependent on the composition of the ECM in this region for the resident stem cells to determine their fate. However, these properties change considerably during neuronal malignancies.

While it was assumed for a long time that neural tissue development was basically dependent on biochemical factors, it has become well accepted that mechanical cues also play an important role in neuron cell development and function. A variety of complex mechanical stimuli, including tensile, compressive, and shear deformations, act on the axonal endings, some of which play the specific role of detecting such signals as mechanosensors through cellular deformation. The axons that belong to the neurons in the PNS lie principally outside the vertebral column and hence experience the constant dynamic mechanical environment in this region. Peripheral neural tissue also experiences tensile deformation during joint movement. The process of differentiation and maturation of neurons is highly dependent on the intricate dynamics of the interaction between the cells and their microenvironment as well as the delicate conformal changes in the cytoskeleton to bring about migration and neurogenesis. Any alteration in the mechanotransduction process or changes in the stiffness of the brain tissue results in aberrant axonal growth and migration. Thus although not understood well earlier, the critical role of mechanical stimuli in neural tissue development and maturation in only being recently recognized.

5.7.2 Mechanotransduction in nerve tissue

The individual components and the intricate cross-linking present in the ECM are responsible for the mechanical and structural characteristics of the brain tissue. The process of neuronal development is characterized by a strong interaction between the integrins and its microenvironment and the ECM of the neuronal tissue. In contrast, the lack of organization in the ECM and changes in its composition are intimately associated with diseases of the brain. Integrin-mediated processes are quite involved in the development and functioning of neuronal cells. A distinct scaffold called the perineuronal net surrounds the neuron and its processes and is very important for the synaptic structure and for regulating synaptic plasticity. It must be noted that the ECM of the CNS is quite different from the ECM of other tissues. This CNS tissue is generally softer than other tissues, although its stiffness changes with aging and disease, which include Alzheimer's disease, multiple sclerosis, etc. As can be deducted from its softness, brain tissue has a high hyaluronic acid content and a large concentration of GAGs and glycoproteins. Even though collagen type IV is present in the brain tissue, it is nonfibrillary, and its content is rather low, highlighting the reason for the low stiffness of this tissue. Fractones, which are structures rich in heparan sulfate and laminin, are found in the vicinity of the blood vessels and serve as neural stem cell niches. Fig. 5.13 is a clear illustration of the neuronal tissue and a comparison of the stiffness of different body tissues.

The fate of stem cells and their differentiation into neurons strongly depend on the stiffness of the neural tissue. A stiffness of <1 kPa, and in general, a softer tissue, is preferred for direction toward a neuronal fate. From a structural perspective, micro/nano topographical features are highly influential in providing contact guidance in the alignment of cell polarity and the outgrowth of axons in a specific direction, attributed to the confinement and alignment of focal adhesions. Beyond just providing directionality, nanotopographical features also appear to favor stem cell differentiation into neurites. Although the mechanism and reason are unclear at the present time, the mere presence of nanotopographical features with specific dimensions on the substrate influences decision-making and fate of neuronal cells. Another manner of directing the preference of stem cells toward neuronal differentiation is to decrease the size of integrin adhesion complexes (IACs) to below that of the focal adhesion complexes by controlling the

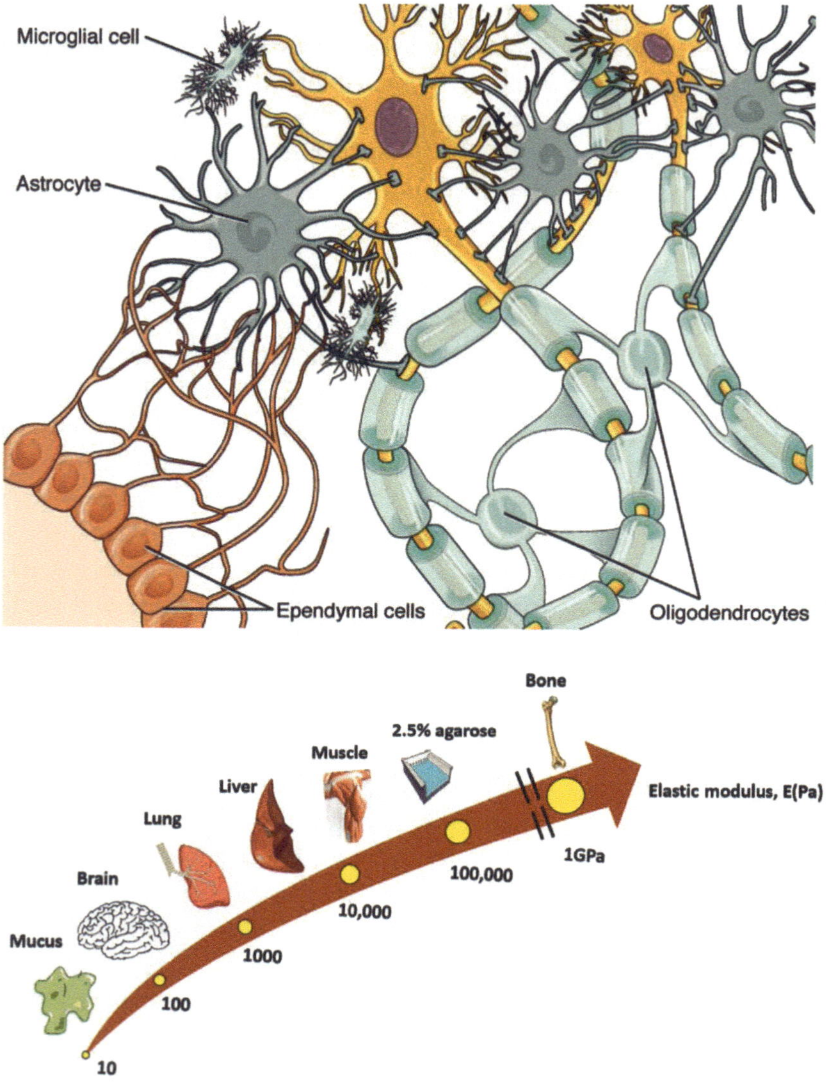

Figure 5.13 Neural tissue of the central nervous system. Illustration of the neural tissue of the brain. The mechanical properties of the constituents of the brain contribute to the lower stiffness of this tissue. The figure below is a comparative chart of the stiffness of various tissues in the body. *From Barnes JM, Przybyla L, Weaver VM. Tissue mechanics regulate brain development, homeostasis and disease. J Cell Sci 2017;130(1):71–82.*

nanotopographical features of the substrates. This results in decreased stress fiber formation and cell rigidity, encouraging neuronal differentiation. Neurite growth cones have a low elastic modulus (~100 Pa) and generate weak forces regulated by processes involving mechanotransduction.

The mechanical processes are extremely important in synaptogenesis, with postsynaptic terminals being rich in cell adhesion molecules, which are mechanosensitive components. The neural circuitry and synaptic connectivity are also mechanosensitive and strongly influenced by ECM constitution and configuration as well as its interaction with integrin. Generally, the brain tissue has mechanical properties that are unique and specifically adapted for neural stem cell behavior and development of this brain tissue. Clearly, any deviation in the properties of the ECM and any disruption to the mechanotransduction signaling will lead to neurodegenerative diseases, including brain cancer. For example, the activation of focal adhesion kinase activates the Rho-associated protein kinase cascade. These proteins, in turn, activate MAP kinases such as extracellular sinal-regulated kinases and YAP, which increase cellular tension in the cytoskeleton. This results in the cell producing more ECM and remodeling it by stiffening the microenvironment at the cell membrane/ECM interface, enhancing the mechanosensitivity of the cell. The fine balance of this mechanocircuitry can be perturbed by even very small changes in the mechanical signaling, causing disease in the brain tissue.

There is no doubt that many mechanical stimuli act on tissues to actuate various physiological functions in the body while these responses at the tissue level are, in turn, responsible for applying mechanical forces to maintain homeostasis in these tissues. Any malfunction of these mechanotransduction processes can lead to pathologies and disease, which can be corrected by suitable mechanotherapies and other remedial actions.

Review and Project

1. Enumerate five characteristics of a mechanical signal as applied to a tissue.
2. What is the effect of compressive force on bone tissue? What are the possible responses to applied shear stress on epithelial cells of the vasculature?
3. Develop a hypothetical cell response map for applied strain as a function of time for fibroblasts in the pulmonary tissue.
4. Generate a hypothetical cell response map for shear stress versus frequency for osteocytes.

CHAPTER 6

Mechanotherapy

6.1 Introduction

The term *mechanotherapy* was first coined in the late 19th century and defined as "the employment of mechanical means for the cure of disease." With time, the term evolved to refer to physical therapies and more specifically exercise therapeutics for injured or diseased tissue in the musculoskeletal system, with a heavy focus on rehabilitation of patients after trauma. The ultimate objective of all physical therapies, including mechanotherapy, is to restore full functionality to the diseased or damaged tissue. Recently, mechanotherapy has been redefined to describe "the employment of mechanotransduction for the stimulation of tissue repair and remodeling." With the evolution of the definition of this term, emphasis has shifted from a mere therapeutic outlook for rehabilitation to a much more scientific understanding of the use of a molecular and cellular basis for treating disease using a plethora of practical medical techniques. The most recent definition, therefore, is that mechanotherapy refers to "any intervention that introduces mechanical forces with the goal of altering molecular pathways and inducing a cellular response that enhances tissue growth, modeling, remodeling or repair." Thus although the mechanical stimulus is applied at the tissue level, the aim is to transmit these forces to elicit a response at the cellular scale and consequently a molecular response to restore functionality to the tissue being rehabilitated. Fig. 6.1 is a simple illustration of this concept of mechanotherapy based on the most recent definition given earlier. The mechanical stimulus is applied at the tissue scale, but this effect is transmitted down to the cellular and molecular scales where actual change takes place to affect the mechanotransduction pathways. The collective response at the cellular scale is manifested as the repair or restoration of the tissue. It is possible that even from the mechanical perspective, the cytoskeletal response from the cell may be purely biomechanical, which regulates molecular-level responses and alter extracellular matrix (ECM) characteristics. This feedback is provided to the diseased or damaged tissue to accomplish repair and restoration.

In mechanotherapy, the mechanical stimulus is applied directly to the tissue. The cells constituting the tissue receive this stimulus based on the mechanical properties of the tissue and undergo the process of mechanotransduction through the signaling pathways while experiencing cytoskeletal rearrangement. The eventual cell response is fed back to the tissue, which then undergoes the necessary modification to accomplish the physiological function. In this overall process, the molecular response could also result in ECM synthesis or degradation and alters the tissue resulting in repair or restoration.

Mechanotransduction
DOI: https://doi.org/10.1016/B978-0-12-817882-9.00007-0

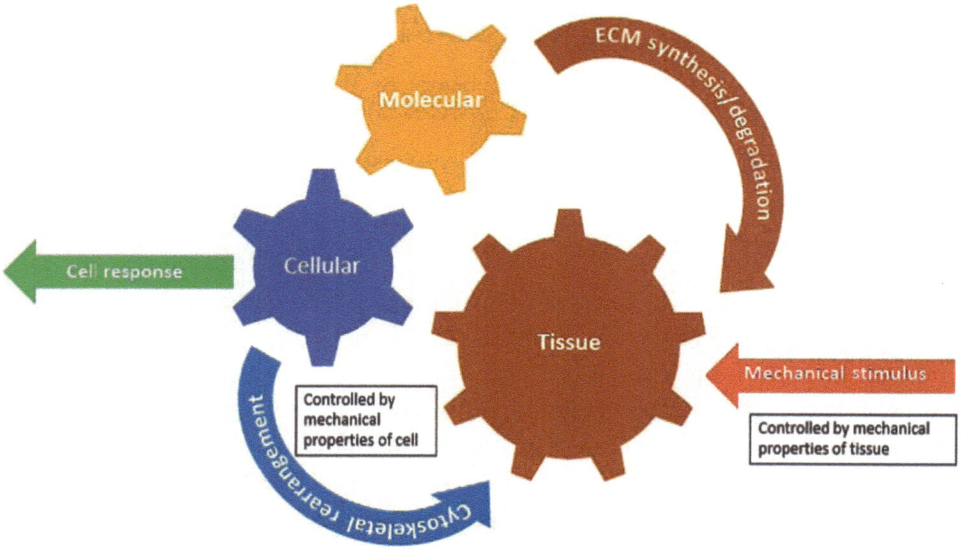

Figure 6.1 Hierarchy of the application of mechanotherapy and the resulting effects. *ECM*, Extracellular matrix.

6.2 Mechanotransduction and disease

Many of the disease characteristics that are observed today are a result of alterations in the physical structure of normal tissues. Many such changes lead to a severely decreased quality of life for the patient, and current solutions are either absent or the challenges resulting from the disease are not appropriately addressed because of a lack of recognition of their physical origin. This could be a result of the fact that the real problem might be at the cellular or molecular scale, which is not easily discernible from examining the patient. However, recently, there is a growing awareness of the relationship between disease and dysfunction and the temporal and spatial responses of cells, especially under mechanical loads. Moreover, current research is focused on harnessing and manipulating these mechanical stimuli to revert disease in some manner and promote healing. Recently, scientists are discovering and understanding the regulatory role adopted by many proteins in the cell under the guidance of mechanical forces. The critical functioning of these molecules in cell physiology and the possibility of disease due to their malfunction in signaling pathways are slowly being accepted based on scientific evidence. There appears to be a physical and mechanical basis for many diseases such as osteoporosis, arthritis, scoliosis, and coronary problems. Many of these disabilities are definitely associated with inappropriate or deficient mechanotransduction and possibly a lack of or excessive mechanical stimuli.

Even in embryogenesis and developmental biology, the critical role played by mechanical forces in the development of many of the organs and their continued maintenance because of mechanical stimuli must be recognized. The importance of mechanics and mechanical forces at the cellular and molecular scales in therapeutic procedures and in innovative applications for the treatment of various diseases is definitely being recognized and considered in recent clinical therapeutics. Considering that human systems are complex hierarchical structures, any mechanotherapy at the organ level translates to be effective at the molecular and cellular scales (Fig. 6.1). Thus understanding the connections that exist between these different hierarchical scales and the corresponding signal transmission paths become important. A tensegrity model has been proposed fairly recently where all the elements in the cell, especially the cytoskeleton, are under a stress equilibrium as a result of the cell applying a contractile force on the substrate to which it is attached at the microscale. When considering tensegrity at the tissue level, this adds up to a rather complicated state of stress, where regardless of the number and type of cells in the tissue, all the cells must act together in coordination to maintain tissue homeostasis. The same is true at the next higher organ level, and any physiological function that must be realized should be aptly coordinated. Obviously, physically modeling such a process while keeping this hierarchical structure in mind for the purpose of mechanotherapy is quite complex. From a simple perspective, any disruption in the stress equilibrium and transmission of force in the mechanical connection between the ECM, the cytoskeletal elements involved, and the nucleus is detrimental to the health of the cell or corresponding tissue and will directly lead to disease. Furthermore, as many of the cellular physiological processes involve the transmission of one or many mechanical signals, even a slight discrepancy in the fidelity of these signals will lead to gene mutations, causing structural changes in the molecular proteins, which are synthesized, which can again result in or lead to a diseased state of the cell or tissue unless the system can repair these small perturbances in the adequate time.

One should not forget the importance of mechanotransduction, as it relates to developmental pathways or in the small force magnitudes experienced by a small group of cells that is tantamount to maintaining tissue homeostasis at a relatively much larger scale. One striking example is maintaining balance resulting from the dynamic equilibrium in the displacements of the stereocilia of hair cells in the ear. The small stresses experienced by vascular smooth muscle cells (VSMCs) with dynamic variations in the blood pressure and the critical requirement to maintain blood flow in the body at the much larger scale are other examples of the significance of small-scale stresses on a localized group of cells working to keep the body in a healthy state. The sense of touch and how this mechanical signal is transmitted to the brain are yet other examples of signal amplification initially detected by a small group of cells.

Mechanical effects at the cellular and molecular levels also have a serious effect on disease progression. Although degenerative issues in the brain are associated with problems in protein folding, alterations in the ECM composition and the corresponding mechanical changes appear to be responsible for the early onset of Alzheimer's and

Parkinson's diseases. Also, changes in the forces sensed by brain cells seem to have a drastic effect on brain function. Another change in the mechanical properties of the brain tissue is caused by traumatic brain injury (TBI). TBI upregulates Tenascin-C in certain regions of the lesion site and causes the stiffening of the tissue, which in the long run is expected to have a negative effect on brain structure and function. Another important effect is that the possible gene mutations that are associated with mechanical stretch in cardiomyocytes influence the response to this stimulus and represent a basis for cardiac disease. Recent studies have shown a clear correspondence in the deficiencies in the stretching mechanisms and heart failure.

The connection between applied mechanical stimulus and the etiology of various diseases is recognized more and more as these relationships become obvious through research and experimental observations. The abnormal responses of tissues and organs to applied mechanical stimuli are critical channels for the development of diseases. Recently, increasing evidence has been gathered, which points to the deviations in brain ECM and neuronal mechanotransductive processes as being the cause of various neuronal pathologies including Alzheimer's disease, schizophrenia, autism, and glioblastoma. Fig. 6.2 is an illustrative concept of how changes in any of the phases of the mechanotransduction process can lead to the development of disease. Here, alterations in cell characteristics and function because of defects in the mechanical environment,

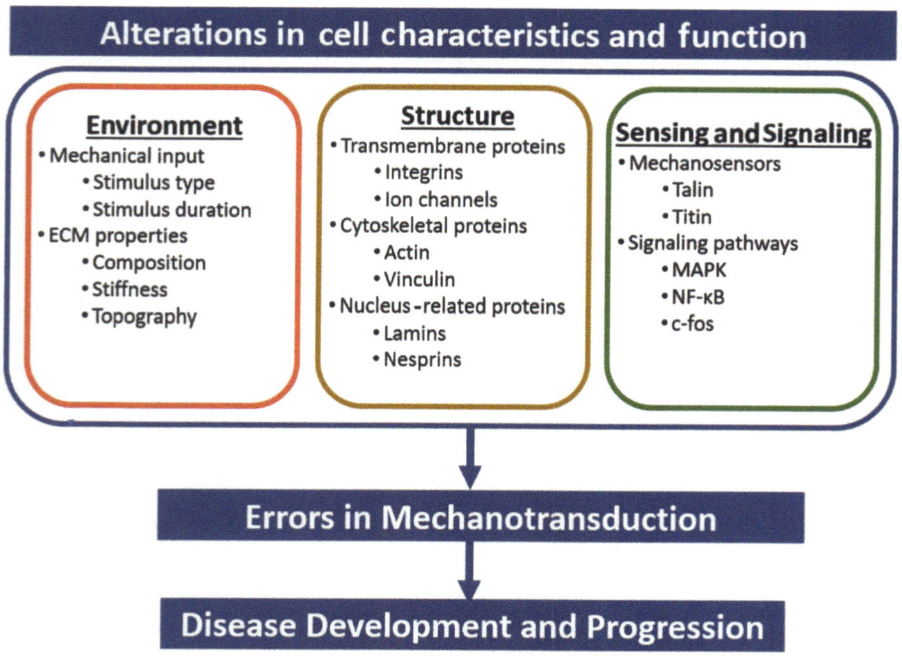

Figure 6.2 Deficiencies in mechanotransduction leads to disease. *ECM*, Extracellular matrix; *MAPK*, mitogen-activated protein kinase; *NF-κB*, nuclear factor kappa B.

including ECM properties, deficiencies on the structure and function of various cellular proteins, and lack of appropriate transmission of signals in the molecular proteins or signaling pathways involved, will result in disease development and progression.

It is worth mentioning the recent understanding that even cancer can be associated with defects in the mechanical characteristics at the cellular scale. The fact that the tumor microenvironment is constantly changing in mechanical characteristics is well known from medical diagnosis. Elevated fluid pressure in the tumor, cell compression, increased stiffness of the ECM, and cellular contractility are well-known observations of tumor tissue. Cancer cells experience a wide range of mechanical forces during metastasis, which enables them to adapt and acquire resistance to successfully colonize the metastatic cell. While the tumor microenvironment is increasing in stiffness, the metastatic cell is becoming more and more compliant with tumor progression to attain properties appropriate to detach from the primary tumor, invade through basement membranes and the ECM by intravasation, to then travel through the capillary or lymphatic vessels to finally attach at a distant site following extravasation to form the secondary tumor.

Disease development and progression occur by any changes in the normal mechanotransduction process during any of the phases in the extracellular environment, structure, and organization of the cell and during cellular sensing and signaling.

Table 6.1 is by no means an exhaustive list of the influence of mechanical force or stretch on various diseases and the implications of the corresponding tissues affected in clinical medicine. However, it highlights the vast influence of mechanical forces in various physiological functions and the universal role in which deficiencies in the mechanotransduction process play on the generation of disease. The importance of the connection between the processes of mechanotransduction and disease development is being generally accepted in the scientific community, although this relationship may not be clearly seen in the data provided in this table. Deficiencies or even alterations in the mechanotransduction process are deemed to contribute to various musculoskeletal disorders, including osteoporosis and muscular dystrophies. Mechanotransduction is also important in the regulation of diverse cellular functions, including protein synthesis, proliferation, migration, and apoptosis. Defects in these processes as a result of irregular mechanotransduction will result in diseases. Table 6.1 highlights some of the diseases associated with deficiencies in the mechanotransduction process and the primary tissues that are affected. For example, the stereocilia in the hair cells are subjected to shear, which results in the temporary opening of the K^+ channels by the tension applied to the stereocilia. Defects in the protein structures involved will not allow restoration of the original shape of the K^+ channels and will result in deafness. Disturbances in the fluid flow at bifurcations in the vasculature will lead to remodeling of the blood vessels and lead to atherosclerosis. The effects of microgravity and lack of exercise with aging will cause deficiencies in the generation of bone tissue and lead to osteoporosis.

Table 6.1 Possible diseases in the various organs/systems resulting from irregularities in mechanoregulation in the human body.

System	Disease	Primary cells/tissue affected
Cardiovascular	Angina pectoris	Myocardial contractile cells
	Atherosclerosis	Endothelial and smooth muscle cells
	Coronary heart disease	Endothelial cells
	Infarction	Cardiomyocytes
	Valve failure	Valvular endothelial and interstitial cells
Gastrointestinal	Constipation	Glandular epithelial tissue
	Diverticulitis	Mucosal and submucosal tissue
	Hemorrhoids	Hemorrhoidal tissue
	Irritable bowel syndrome	Intestinal epithelial cells
Musculoskeletal	Arthritis	Articular cartilage/chondrocytes
	Osteopenia	Osteoblasts
	Osteoporosis	Osteocytes
	Scoliosis	Bone tissue
	Spondylitis	Cervical vertebral tissue
Nervous	Alzheimer's disease	Neurons and microglia
	Glioblastoma	Glial cells
	Hydrocephalus	Ependymal cells
	Stroke	Neurons
	Traumatic brain injury	Multiple cells in the brain
Pulmonary	Asthma	Alveolar tissue
	Emphysema	Airway epithelial cells
	Fibrosis	Pleural tissue
	Embolism	Pleural tissue
	Edema	Pleural tissue
Others	Glaucoma	Optic neurons
	Deafness	Hair cells
	Polycystic kidney disease	Epithelial cells
	Tumors	Multiple cell types and tissues

The correspondence between disease and irregularities in mechanoregulation may be sudden or it may be gradual as the develops over a period with a growing loss in mechanical function. The unraveling of the mechanisms involved in mechanotransduction is critical to the current concepts of mechanotherapy. Hence, understanding the molecular pathways of mechanotransduction will play a key role in not only identifying target molecules for therapeutic interventions but also stimulation methods to aid tissue regeneration using mechanotherapy. Recent emphasis has been placed on understanding the role of integrin-regulated mechanisms that influence various cellular functions, including the internal residual forces in the cytoskeleton and cell–matrix interactions. Other less important cell structures that are implicated in the transmission of applied mechanical stimuli include adherens junctions, gap junctions, and primary cilia.

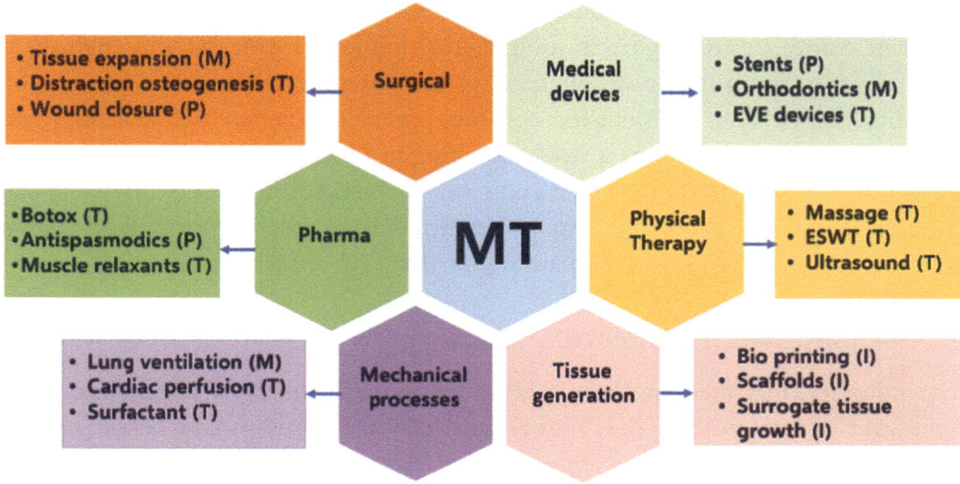

Figure 6.3 Mechanotherapy (MT) options that are clinically available and are based on the use of medical devices, mechanical processes, pharmaceuticals, physical therapy, surgery, and tissue generation with the purpose of altering mechanical characteristics of tissues and cells. Also included is a grouping of the MTs based on whether they are mechanics dependent (M), target tissue dependent (T), purpose dependent (P), or interventional-level (tissue, cellular, or molecular) dependent (I). *ESWT*, Extracorporeal shockwave therapy.

Based on our understanding of the type of molecules involved in mechanotransduction and their degree of involvement in disease-related processes, a variety of therapies can be developed to combat these diseases and improve the quality of life for the patient. Mechanotherapy is normally administered at the tissue level where there is easy and direct access to the site where therapy is required. However, with the recent change in direction to focus mechanotherapy at the cellular and molecular levels, this might involve invasive treatment or injectables. Fig. 6.3 provides a palette of possible mechanotherapies without actually indicating the specific molecular or cellular pathways that are utilized to accomplish tissue repair or remodeling. As a simple example, stent placement in blocked coronary arteries is a clinical mechanotherapy that is readily available today, although it involves an invasive treatment. Another more complex mechanotherapy is distraction osteogenesis where a tensile load is applied on bone to enable it to grow by the generating tissue. In cases of infections, which are difficult to treat, vacuum therapy uses mechanical forces to assist wound closure. At the cellular scale, the process of mechanotransduction indicates the initiation through some sort of mechanical coupling, which is then translated into the cell either by intracellular or cell-to-cell communication to affect a response from the cell or group of cells. This may include injecting a muscle relaxant or ingestion of antispasmodics for therapeutic purposes. Other areas that may not have been considered earlier as involving mechanical forces such as tissue expansion, which is a mechanics-dependent therapy, or cardiac perfusion, which is a target-dependent procedure, are now being accepted as mechanotherapies. Surrogate tissue growth using animal models to generate and engineer tissue for further implantation in humans is also another therapy that can involve mechanical forces. Thus, in a macroscopic sense, mechanotherapy is a process where a mechanical trigger is applied to a tissue, resulting in this force signal being transmitted to the cells to bring about the desired end result by manipulating molecular-scale pathways. Fig. 6.3 also provides a corresponding grouping of mechanotherapies depending on the type of mechanics, the target tissues, or the level of mechanointerventions.

6.3 Mechanotherapy: Molecular scale

If one considers that mechanotherapy at the molecular scale is viable for tissue regeneration and disease control, it is only logical that the focus should be on not only understanding the molecular pathways of mechanotransduction but also on the research methods to influence and manipulate these molecular targets appropriately. Numerous potential targets are available for therapeutic interventions at the molecular level. For example, intracellular calcium ion transport, regulation of prostaglandin (PGE2) and NO, and Wnt/β-catenin signaling are all intimately involved in bone regeneration. Enhancing or inhibiting certain signaling molecules using mechanical stimuli will result in the necessary therapeutic intervention for regeneration of tissue after osteoporosis. Similarly, appropriate control of mechanotransduction signaling molecules or processes such as integrin or MAPK pathway using mechanotherapy may be effective in the treatment of benign or malignant fibroproliferative disorders.

In application of the correct mechanotherapy, any one of the four phases in the mechanotransduction process (refer to Fig. 3.2) becomes a viable target: (1) mechanocoupling, where the applied external mechanical stimulus is adapted into a mechanical signal in the cell vicinity; (2) biochemical coupling, where the local mechanical signal is transduced through one or various biochemical reactions for protein synthesis; (3) signal transmission from the sensor cell to the effector cell; and finally, (4) the effector cell response. Various molecular mechanosensors in the intracellular regime are prime targets for mechanotherapy through mechanotransduction. Recent discoveries that integrins are mechanosensors and have potential as target molecules to participate in mechanotherapy have paved the way for studying structural relationships between these and other mechanosensors such as gap junctions, primary cilia, etc., with the goal of influencing the latter sensing structures through the manipulation of integrin. One interesting example is the possibility of providing mechanotherapy against bone loss from immobilization by targeting the gap junctions, which allow for the passage of PGE2 through the hemichannels formed by connexins. The presence of primary cilia has been highlighted as critical to the mechanosensory process for the migration of VSMCs in wound healing and hence a viable therapeutic target.

6.4 Mechanotherapy: Tissue level

The principal objectives of mechanotherapy at the tissue level is repair and restoration, for example, enhancing wound healing. Again, although the therapy is administered at the tissue level, a number of health benefits at the microscale are observed. In what follows, some mechanotherapies currently employed at the tissue level are briefly explained. An important therapeutic intervention is microdeformational wound therapy, which entails the application of a vacuum to the surface of the wound with the objective of withdrawing extracellular fluid, stabilizing the wound environment, and encouraging wound closure. The procedure also promotes neovascularization caused by the mechanical stimulus either by angiogenesis or expansion of blood vessels present in the wound region. This improvement in wound healing is adjudicated to the upregulation of vascular endothelial growth factor proteins and the increased gene expression of interleukin 8 and CXCL5, two chemoattractants for leukocytes directly resulting from the applied mechanical stimulus. Soft tissue expansion therapy is a procedure applied to overstretch the skin to generate new skin. This tissue generation is also a result of the applied mechanical stimulus, which increases cell proliferation and neovascularization. To attain such tissue generation, epidermal growth factors are upregulated by the force applied at the tissue level, which, at the molecular level, elicits changes in cell function. A very important technique for tissue generation at the bone level using mechanical forces is distraction osteogenesis. This has been widely utilized to induce bone tissue regeneration to correct congenital bone deformities, nonunion fractures, and other bone defects by direct membranous ossification across the distraction gap. With time, this gap is bridged by new tissue filling in the region of the bone defect. Distraction osteogenesis can be multidirectional and has been successfully employed in complex craniofacial reconstruction. From the point of view of mechanotransduction, it appears that small strains favor bone formation, whereas larger values of strain encourage fibrous tissue. This technique has been accompanied by the expression of c-Src and bone morphogenetic protein, indicating the role of integrin-mediated signaling in this process.

A number of other mechanotherapeutic techniques are available, including hyperbaric oxygen treatment to treat elevated pressure and hyperoxia. To connect to the molecular scale, it may be useful to know that transforming growth factor-β and VEGF expressions are increased as a result of the treatment. Lung protective and cardiac offloading procedures have been adopted as part of mechanotherapy to reduce stress on these organ-related tissues to protect from harm due to excessive loads. Cardiac offloading has resulted in increased oxygen supply and efficient blood perfusion in myocardial tissue following infarction. This has been related to the restoration of local Ca^{2+} homeostasis and reversal of pathological tubule remodeling.

6.5 Emerging trends in mechanotherapy

It may not be always physically feasible to provide large-scale mechanical therapies to patients recovering from trauma or disease because of their inherent inability to bear such forces. In such cases, novel methods to focus the therapy to more localized tissue regions may solve the problem at hand. For example, it is difficult to apply a physical deformation to the bone matrix directly and transmit this force to osteocytes, which are the mechanosensors in bone tissue. Hence, direct loading of the intramedullary cavity to induce interstitial fluid flow without unnecessarily loading the tissue would be an excellent therapeutic option. This may be accomplished by bending the bone locally through small, intermittent stimuli. It must be recognized that the intramedullary cavity also contains mesenchymal stem cells (MSCs) and hematopoietic cells, which will clearly be influenced by this localized therapy and probably contribute to this tissue regenerative process. One technique that is currently being developed in this direction is low-intensity vibration (LIV), which can stimulate cells in bone and muscle without applying large-scale deformation to the tissue. The manner in which LIV induces regeneration at the cellular level is currently not well understood. However, this is very much in line with *in vitro* observations where pulsatile low magnitude forces are very effective in the growth of bone cells and controlling the fate of MSCs to bone tissue. Another novel technique to externally apply stimulus to bone tissue is low-intensity pulsed ultrasound and appears to work both for osteogenesis and chondrogenesis. Again, the magnitude of the applied mechanical stimulus will be effective in a more selective therapeutic treatment. Tendon, when mechanically loaded, shows upregulation of insulin-like growth factor, which enhances cell proliferation and matrix remodeling, whereas overloading muscle results in a local increase in mechanogrowth factor, which leads to hypertrophy. Controlled loading of muscle tissue after injury allows for faster growth and alignment of regenerating myotubes. Continuous passive motion of articular cartilage populated with mechanosensitive chondrocytes also improves patient comfort after injury to the patella, while fracture healing is accelerated under intermittent compressive mechanotherapy.

There is no doubt regarding the effectiveness of mechanics-based therapies for the musculoskeletal system. In this case, nearly every interventional therapy involves some kind of mechanical force, either externally or intrinsically. However, recent efforts in mechanotherapy aim to extend these interventions beyond the musculoskeletal realm to other tissues and organs. Other challenges exist in translating mechanotherapy applied at the tissue level to mechanotransduction at the cellular scale. While it is easy to surmise that any therapeutic intervention using mechanical stimuli will produce an effect at the molecular level, the specificity of the applied intervention to a particular cell type is debatable. In a tissue, composed of many cell types, focusing on a specific cell type without affecting other cells may be nearly impossible. Also, thresholds with regard to the type of force, magnitude, stimulus duration, and timing become critical factors for providing effective and focused mechanotherapy. This is complicated by the significant cross-talk between the various molecular pathways that may be directly affected by the therapy provided. At this point, further research and study are necessary to obtain a knowledgeable grasp on both scientific concepts and practical applications, although there is little doubt that the mechanotherapy route incorporating an in-depth understanding of mechanotransduction of molecular-scale mechanisms will become clinically relevant in the near future.

Another modality in vogue is to combine mechanotherapy with a biologically active agent. For example, mechanotherapy combined with the intermittent introduction of parathyroid hormone (PTH) encourages adaptive bone anabolic response. The PTH stimulates osteocytes and osteoblasts when introduced intermittently and when the mechanotherapy is provided at the right period after PTH is administered. Myostatin, which is a negative regulator of adult skeletal muscle mass, when inhibited through other pharmacological agents and combined with appropriate mechanotherapy, is effective in synthesizing muscle mass.

6.6 Cancer therapy and mechanotransduction

As discussed in an earlier chapter, there is a plethora of evidence relating to the effect of mechanical stimuli and the process of tumor growth and metastasis. Based on these, there are three target areas for possible cancer therapy:

(a) *Cell contractility*. An increase in actomyosin contractility is concordant with ECM stiffening, which is conducive to the growth of tumors. Any therapy that can regulate the mechanotransduction mechanism and result in a decrease in ECM stiffness will be positive for cancer treatment.

(b) *Solid stress*. Increased stress is placed on adjacent tissue because of tumor growth and stiffening. This exerts increased pressure on the vasculature impeding efficient drug delivery and leading to hypoxia. Any therapy to decrease this solid stress will also be beneficial by providing more efficient treatment against the tumor.

(c) *ECM stiffness*. An upregulation in collagen deposition can cause fibrosis in the tumor region and increase ECM stiffness. This results in faster growth of the tumor resulting in strong implications for metastasis.

Ameliorating the stress and stiffness of the tumors will increase the effectiveness of cancer therapy and decrease the possibility of metastasis.

6.7 Other therapeutic effects of mechanical stimulus

While clearly therapy is treatment applied with the goal of healing and restoration of tissue function in mind, a number of other effects occur because of inadvertent mechanical stimulus. Physical forces are clearly important during fetal lung development, metabolism of surfactants, and mechanical ventilation. The filling of liquid in the lung and fetal breathing movements are critical for lung growth. Any discrepancy in this pattern results in disruption to proper lung growth and results in disease. A better understanding of continual, repetitive distention of the lung tissues will lead to appropriate therapies to prevent lung hypoplasia and enhance the maturation of this organ. The intermittent deep breaths during the respiratory process are important for surfactant secretion, although details of this mechanism and its effect on surfactant activity has not yet been clearly deciphered.

While many of the effects of mechanical stimuli are positive, negative results are also possible. In a negative sense, deficiencies in the mechanotransduction process can also cause the upregulation of collagen type I during the cyclic distention process leading to fibrosis. For example, direct strain or high shear on lung tissue caused by high alveolar pressure are deemed to cause injury to the lung. Interstitial fibrosis and local differences in lung tissue compliance also result in disease. Elastic recoil of the lung tissue is affected over time as a result of smoking, as also gas exchange in the lung alveoli. Emphysema results as a result of cigarette smoking, environmental pollutants, and genetic factors, and this results in a decline in effective lung function. Long periods of physical inactivity can cause dystrophy of muscles and osteoporosis as a direct result of mechanical unloading. As mentioned earlier, the need for mechanotherapy is in itself an indirect reference to the pervasive nature of the mechanotransduction process not only in our daily physical activities but so much more at the cellular and molecular scale. There is thus a critical need to understand this process to prescribe effective mechanotherapies for intervention to repair and restore diseased tissues.

Review and Project

1. Enumerate three mechanotherapy options appropriate for bone tissue. Describe the possible effects of such therapies on the cellular and molecular scales.
2. Review the different mechanotherapies mentioned in this chapter. Describe an effective therapeutic treatment for the ablation of small benign tumors on the epidermis?
3. The lung capacity of a person has to be increased for more oxygen intake. Can you develop an appropriate mechanotherapy to accomplish this?
4. Can you prescribe a novel effective mechanotherapy for the treatment of osteoporosis?

CHAPTER 7

Future thrusts in the field of mechanotransduction

Although our understanding of mechanotransduction has increased over the last two decades, many of the mechanisms underlying the process itself remain poorly defined or our knowledge about the fine details of this phenomenon is fragmented at best. The paucity of scientific rationale in explaining the various empirical observations opens new avenues for research in the area of mechanotransduction. Based on the current state-of-the-art, possible topics or areas where further research is necessary are suggested. Although a complete list of research topics is obviously not provided, in the following sections, the focus will be on some open questions regarding signaling mechanisms, cell—extracellular matrix (ECM) interaction, stem cells, tissue mechanotransduction, and mechanotherapy.

The critical role of many protein molecules in signaling pathways involving mechanotransduction must be studied, and since their malfunction may lead to manifestation of various diseases, it is important to understand these processes clearly to provide remedial solutions for effective therapies. For this, new and innovative experimental techniques must be developed to observe molecular processes inside single cells in greater detail, especially in vivo to deepen understanding and knowledge in this area. Clarity is needed regarding the mechanosensing mechanisms involving transmembranal integrins and the role of these proteins in mechanotransduction. Knowledge is still lacking in the initiation of mechanochemical transduction and the interaction between the biomechanical and biochemical signaling pathways even though these pathways themselves are fairly well identified. Identification of the mechanoresponsive genes may be essential for understanding this phenomenon better. Several mechanotransduction pathways have been identified in vitro although their relevance in physiology in vivo is still unknown. On the other hand, other functions, both at cell and tissue levels, which are regulated by mechanical forces, are still lacking a detailed mechanical explanation at the cell molecular scale. While it is now known that the same group of cells can experience various types of stress and respond specifically to the particular mechanical signal, understanding the signaling pathways to regulate them to take advantage of the specificity of their responses will involve manipulating genetic mechanisms by focused application of mechanical signals with emphasis on the duration and magnitude of these forces. The sensory mechanism by which myocytes regulate mechanical forces and the manner in which these are converted to electrochemical signals and transferred to the nucleus is still unknown. Also, although a large number of

Mechanotransduction
DOI: https://doi.org/10.1016/B978-0-12-817882-9.00006-9

gene mutations, which cause cardiovascular disease, are known, the underlying mechanisms are still not well understood, and research is needed in this case and for other diseases to determine whether mechanical stimuli will be effective in dealing with disease progression. Different forms of stretch results in the activation of distinct signaling pathways in different cells (endothelial cells, fibroblasts, and cardiomyocytes), although each individual cell type may respond in divergent manners even for the same mechanical stimulus. To understand this complex response and separate specific signaling pathways, each cell type must be analyzed separately, and this remains a challenge. The molecular identity of mechanosensors in cardiomyocytes is still not clearly known.

Understanding of the remodeling of the ECM by different cell types and underlying reasons must be investigated. The mechanism by which applied stress is transferred from the substrate to the nucleus to accomplish gene transcription is still far from being clearly defined. Physical and mathematical modeling of the complete process is needed, as well as modeling of each step to realistically couple each of the stress transfer stages from ECM to integrins to actin stress fiber elements via the focal adhesion protein complex must be thoroughly understood. Considering its versatility and the large number of molecules that compose the ECM, a fundamental study encompassing the role of each of these molecules and the manner in which they affect overall ECM and corresponding cellular activity is imperative. In addition, when the plethora of roles that each of these molecules may play in multiple tissues at specific times of organogenesis is considered, a rich avenue of research is open for an in-depth study of the ECM both in development and disease. It is possible that understanding the causes of any dysregulation in the cross-talk between the ECM and other tissue components may throw light on the etiology of congenital defects and cancers.

While much progress has been made in understanding cell—substrate interactions, there is still a lot of room for deciphering the mechanisms corresponding to this interaction. It is well known that adherent cells form focal adhesions to attach to available substrates. However, the reasoning for the choice of substrate or a particular region of the substrate is not clear. The manner in which cells sense local signals due to substrate elasticity or topography is another area that needs detailed research. Are the connections between the integrin proteins and the fibrous proteins in the ECM chemical in nature, or these just physical cross-links? The driving force for the nucleation and growth of focal adhesions and their assembly and disassembly during cell migrations are other questions that must be answered. While the need for mechanical force in many cell physiology processes is understood, the knowledge of the exact role and influence of the mechanical stimulus in these processes is shallow at best. What is the mathematical relationship between applied force, stretch, or shear stress resulting from fluid flow and the nucleation of focal adhesions? In what manner does the magnitude of this force or stretch mathematically relate to the number of focal adhesions, their size distribution, or their density? Again, does this vary with normal cell type? Is this different in cancer cells? How about stem cells? Does cell migration occur through autocrine or paracrine sensing processes involving contractile cell stresses? Is the migration of cancer cells controlled by mechanotransduction? Is there a threshold force level that has to be applied on the substrate for cells to initiate migration?

A highly regulated ECM is critical for tissue homeostasis and controlled cellular processes. Any perturbation in this balance will result in pathological conditions, which can result in disease progression and death. Clearly understanding the cause of these perturbations in the ECM homeostasis and the changes occurring in the ECM composition as a result is imperative for developing novel and innovative therapeutic agents to combat a variety of diseases. Continuously monitoring the ECM may ultimately become an established clinical practice in the effective treatment of disease. While it is generally accepted that any disruptions to the mechanical signaling scheme do lead to deficiencies in tissue function and hence disease, there is an emerging theme that such changes in the ECM microenvironment also lead to positive feedback. The protumorigenic mechanosignaling feedback that is received does also result in resistance to therapy and inadequate prognosis. This feedback circuit needs to be examined further to locate new biomarkers and target for effective therapy. Although there is much to be researched about the role of mechanical stimulus in neurogenerative diseases, it is now recognized that changes in the physical and mechanical properties of brain tissue are intimately associated with the progression of Alzheimer's disease, Parkinson's disease, and multiple sclerosis. Knowledge of tissue ECM mechanics, fluid flow, and mechanosensitive ion channels in synchrony with soluble factors in the cell migrations would be useful in regulating growth and differentiation in both embryonic and adult brains.

It is now understood that even though mechanical stimulus alone may accomplish stem cell differentiation, the influence of biochemical factors may enhance this process. Therefore it may be useful to develop maps with regimes that highlight conditions that will encourage stem cell differentiation. For example, for mesenchymal stem cells (MSCs) in the presence of a certain growth factor such as BMPs or a culture medium with dexamethasone, a 2D map of mechanical load versus frequency whose combinations can encourage MSC differentiation can be developed. Alternately, stretch versus time duration map for the same cells under similar conditions can provide experimental data under which differentiation can be effectively accomplished (a concept map along these lines is provided in Fig. 5.8). These maps may be very useful in designing mechanotherapies with exercise regimens for patient to stimulate either MSC-based implants or stem cells in tissue niches in the body. Generally, the duration of stimulus continues to be a challenge for effective differentiation as well as the certainty in the determination of stem cell fate. Simulating stem cell behavior using computer modeling and the use of artificial intelligence to decide stem cell fate are important future directions for research.

Although the concept of a niche is well known in stem cell biology, various aspects of the microenvironment in these niches are still not well known. The role of the ECM in these niches is still unclear and the interaction of the components of the ECM with the stem cells is not entirely understood. Why would the ECM in the stem cell niche favor differentiation while the nonstem cell niche ECM help retain the stem cell properties? The comprehensive signaling mechanisms by which stem cell fate is regulated by the local microenvironment stiffness is still unknown. Also, the range of stiffnesses and how they influence lineage differentiation in MSCs is still moot. The molecular machinery involved in mechanotransduction signaling pathways in both fully differentiated and stem cells is still unclear. Although some fundamental mechanisms for mechanotransduction have been proposed, new mechanistic explanations for observed cell behavior are required including cross-talk between signaling pathways where costimulation is necessary to encourage differentiation of stem cells into a certain specific lineage. Also, the process by which a specific pathway is chosen by individual stem cells, whether for division, self-renewal, or differentiation is currently unknown. Understanding the mechanical environment in the stem cell niches in tissues, which are more prone to mechanical stimulation, and the ability to selectively target these regions may have important implications in functional therapies for the treatment of diseases.

It is extremely important to understand the variation in the biomechanical properties of tissues during embryonic development and equally important to study the changes in normal stiffness of the different tissues with aging and disease. Abnormal tissue stiffness has been observed in fibrosis, and its progression into tumor growth is an important topic of study and research. However, at the molecular and cellular levels, our understanding regarding the transformed role of cells under changing tissue properties is still shallow. Much empirical data are available about substrate characteristics, including stiffness and topography, on cell attachment and migration. However, there is a lack of exactitude regarding the stiffness parameter that is optimum for attachment and differentiation for different cell types. Also, there is as yet no consensus with respect to the appropriate topography that is conducive to cell migration and for tissue generation.

There is an urgent need for implants in a variety of patients. The paucity of viable autologous tissue implants has resulted in a recent spurt in the growth of novel tissue engineering techniques. Avascular tissues such as cartilages present a unique challenge in tissue generation in in vitro bioreactors. A novel route to obtain such tissue is through the use of in vivo bioreactors such as small- or medium-sized animals where MSCs can be injected in appropriate tissues or joint spaces and later subjected to the appropriate mechanical stimuli to provide guidance for these stem cells to differentiate into the desired cell type and possibly grown into tissue. Engineered cartilage and bone show promise for development using the in vivo reactors.

Several areas are available for further research in the field of mechanotransduction. There is much to be researched and even more to be discovered to enable us to understand and apply the principles of mechanotransduction. The author hopes that this book is just a starting point for the reader to explore these and many other topics to contribute to the generation of new knowledge in the field of mechanotransduction.

Bibliography

[1] Ravichandran A, Lim J, Chong MSK, Wen F, Liu Y, Pillay YT, et al. In vitro cyclic compressive loads potentiate early osteogenic events in engineered bone tissue. J Biomed Mater Res B Appl Biomater 2017;105:2366–75.

[2] Altman GH, Horan RL, Martin I, Farhadi J, Stark PR, Volloch V, et al. Cell differentiation by mechanical stress. FASEB J 2002;16:270–2.

[3] Angele P, Schumann D, Angele M, Kinner B, Englert C, Hente R, et al. Cyclic mechanical compression enhances chondrogenesis of mesenchymal progenitor cells in tissue engineering scaffolds. Biorheology 2004;41:335–46.

[4] Bader DL, Knight MM. Biomechanical analysis of structural deformation in living cells. Med Biol Eng Comput 2008;46:951–63.

[5] Baker BM, Chen CS. Deconstructing the third dimension – how 3D culture microenvironments alter cellular cues. J Cell Sci 2012;125:3015–24.

[6] Banes AJ, Tsuzaki M, Yamamoto J, Brigman B, Fischer T, Brown T, et al. Mechanoreception at the cellular level: the detection, interpretation, and diversity of responses to mechanical signals. Biochem Cell Biol 1995;73:349–65.

[7] Barnes JM, Przybyla L, Weaver VM. Tissue mechanics regulate brain development, homeostasis and disease. J Cell Sci 2017;130(1):71–82.

[8] Beachey W. Respiratory care anatomy and physiology: foundations for clinical practice. 4th ed. Elsevier; 2018.

[9] Belin S, Zuloaga KL, Poitelon Y. Influence of mechanical stimuli on Schwann cell biology. Front Cell Neurosci 2017;11:347.

[10] Bello SA, de Jesús-Maldonado I, Rosim-Fachini E, Sundaram PA, Diffoot-Carlo N. In vitro evaluation of human osteoblast adhesion to a thermally oxidized γ-TiAl intermetallic alloy of composition Ti-48Al-2Cr-2Nb (at.%). J Mater Sci Mater Med 2010;21:1739–50.

[11] Boada-Lopez J, DeJesus-Maldonado I, Jerez J, Romañach R, Diffoot-Carlo N, Sundaram P. Collagen abundance in mechanically stimulated osteoblast cultures using near infrared microscopy. J Biomech 2013;46:2442–50.

[12] Brighton T, Fisher Jr JRS, Levine ST, Corsetti JR, Reilly T, Landsman AS, et al. The biochemical pathway mediating the proliferative response of bone cells to a mechanical stimulus. J Bone Joint Surg 1996;78:1337–47.

[13] Campbell HK, Maiers JL, DeMali KA. Interplay between tight junctions & adherens junctions. Exp Cell Res 2017;358:39–44.

[14] Chen YJ, Huang CH, Lee IC, Lee YT, Chen MH, Young TH. Effect of cyclic mechanical stretching on the mRNA expression of tendon/ligament-related and osteoblast-specific genes in human mesenchymal stem cells. Connect Tissue Res 2008;49:7–20.

[15] Chien S, Li S, Shyy JY-J. Effects of mechanical forces on signal transduction and gene expression in endothelial cells. Hypertension 1998;31:162–9.

[16] Chighizola M, Dini T, Lenardi C, Milani P, Podestà A, Schulte C. Mechanotransduction in neuronal cell development and functioning. Biophys Rev 2019;11(5):701–20.

[17] Cora JJ, Diffoot-Carlo N, Sundaram PA. Vinculin expression in MC3T3-E1 cells in response to mechanical stimulus. Data in Brief 2016;6:94–100.

[18] Cox TR, Erler JT. Remodeling and homeostasis of the extracellular matrix: implications for fibrotic diseases and cancer. Dis Models Mech 2011;4(2):165–78.

[19] Cui H, Miao S, Esworthy T, Zhou X, Lee SJ, Liu C, et al. 3D bioprinting for cardiovascular regeneration and pharmacology. Adv Drug Deliv Rev 2018;132:252–69.

[20] Darling EM, Athanasiou KA. Biomechanical strategies for articular cartilage regeneration. Ann Biomed Eng 2003;31:1114–24.

[21] Delaine-Smith RM, Reilly GC. Mesenchymal stem cell responses to mechanical stimuli. Muscles Ligaments Tendons J 2012;2:169–80.

[22] DuFort CC, Paszek MJ, Weaver VM. Balancing forces: architectural control of mechanotransduction. Nat Rev Mol Cell Biol 2011;12(5):308–19.

[23] Dye BR, Miller AJ, Spence JR. How to grow a lung: applying principles of developmental biology to generate lung lineages from human pluripotent stem cells. Curr Pathobiol Rep 2016;4:47–57.

[24] Engler AJ, Sen S, Sweeney HL, Discher DE. Matrix elasticity directs stem cell lineage specification. Cell 2006;126(4):677–89.

[25] Feher J. The cellular basis of cardiac contractility. Quantitative human physiology. 2nd ed. Elsevier; 2012. p. 547–55.

[26] Felsenthal N, Zelzer E. Mechanical regulation of musculoskeletal system development. Dev Adv Dev Biol Stem Cells 2017;144:4271–83.

[27] Finger AR, Sargent CY, Dulaney KO, Bernacki SH, Loboa EG. Differential effects on messenger ribonucleic acid expression by bone marrow-derived human mesenchymal stem cells seeded in agarose constructs due to ramped and steady applications of hydrostatic pressure. Tissue Eng 2007;13:1151–8.

[28] Ford AJ, Rajagopalan P. Extracellular matrix remodeling in 3D: implications in tissue homeostasis and disease progression. Nanomed Nanobiotechnol 2018;10:e1503.

[29] Frantz C, Stewart KM, Weaver VM. The extracellular matrix at a glance. Cell science at a glance. J Cell Sci 2010;123:4195–200.

[30] Franze K. The mechanical control of nervous system development. Dev Adv Dev Stem Cells 2013;140:3069–77.

[31] Garcia CSNB, Prota LFM, Morales MM, Romero PV, Zin WA, Rocco PRM. Understanding the mechanisms of lung mechanical stress. Braz J Med Biol Res 2006;39:697–706.

[32] Geiger B, Spatz JP, Bershadsky AD. Environmental sensing through focal adhesions. Nat Rev Mol Cell Biol 2009;10:21–33.

[33] Ghiasi MS, Chen J, Vaziri A, Rodriguez EK, Nazarian A. Bone fracture healing in mechanobiological modeling: a review of principles and methods. Bone Rep 2017;6:87–100.

[34] Gilbert PM, Havenstrite KL, Magnusson KEG, Sacco A, Leonardi NA, Kraft P, et al. Substrate elasticity regulates skeletal muscle stem cell self-renewal in culture. Science 2010;329(5995):1078–81.

[35] Goetzke R, Sechi A, De Laporte L, Neuss S, Wagner W. Why the impact of mechanical stimuli on stem cells remains a challenge. Cell Mol Life Sci 2018;75:3297–312.

[36] Goetzke R, Franzen J, Ostrowska A, Vogt M, Blaeser A, Klein G, et al. Does soft really matter? Differentiation of induced pluripotent stem cells into mesenchymal stromal cells is not influenced by soft hydrogels. Biomaterials 2018;156:147–58.

[37] Goldmann WH. Mechanosensation: a basic cellular process. Prog Mol Biol Transl Sci 2014;126:75–102.

[38] Haase K, Al-Rekabi Z, Pelling AE. Mechanical cues direct focal adhesion dynamics. Prog Mol Biol Transl Sci 2014;126:103–34.

[39] Herzog W. The biomechanics of muscle contraction: optimizing sport performanceBiomechanik der Muskelkontraktion: Optimierung sportlicher Leistungen. Sport-Orthop-Sport-Traumatol-Sports Orthop Traumatol 2009;25:286–93.

[40] Huang C, Holfeld J, Schaden W, Orgill D, Ogawa R. Mechanotherapy: revisiting physical therapy and recruiting mechanobiology for a new era in medicine. Trends Mol Med 2013;19(9):555–64.

[41] Ingber DE. Mechanobiology and diseases of mechanotransduction. Ann Med 2003;35(8):564–77.

[42] Iskratsch T, Wolfenson H, Sheetz MP. Appreciating force and shape-the rise of mechanotransduction in cell biology. Nat Rev Mol Cell Biol 2014;15(12):825–33.

[43] Jaalouk DE, Lammerding J. Mechanotransduction gone awry. Nat Rev Mol Cell Biol 2009;10:63–73.

[44] Jagodzinski M, Drescher M, Zeichen J, Hankemeier S, Krettek C, Bosch U, et al. Effects of cyclic longitudinal mechanical strain and dexamethasone on osteogenic differentiation of human bone marrow stromal cells. Eur Cell Mater 2004;7:35–41.

[45] Ju W. The nervous system and nervous tissue. In: Neuroscience: Canadian 1st edition open textbook. Creative Commons Attribution 4.0 International License, 2018.
[46] Khan KM, Scott A. A mechanotherapy: how physical therapists' prescription of exercise promotes tissue repair. Br J Sports Med 2009;43(4):247−52.
[47] Kilian KA, Bugarija B, Lahn BT, Mrksich M. Geometric cues for directing the differentiation of mesenchymal stem cells. Proc Natl Acad Sci USA 2010;107(11):4872−7.
[48] Klein-Nulend J, Bacabac RG, Mullender MG. Mechanobiology of bone tissue (mecanobiologie du tissu osseux). Pathol Biol 2005;53:576−80.
[49] Knöll R, Hoshijima M, Chien K. Cardiac mechanotransduction and implications for heart disease. J Mol Med 2003;81(12):750−6.
[50] Ko KS, McCulloch CA. Intercellular mechanotransduction: cellular circuits that coordinate tissue responses to mechanical loading. Biochem Biophys Res Commun 2001;285:1077−83.
[51] Kuo Sh W, Lin HI, Hui-Chun HJ, Shih YRV, Chen HF, Yen TJ, et al. Regulation of the fate of human mesenchymal stem cells by mechanical and stereo-topographical cues provided by silicon nanowires. Biomaterials 2012;33:5013−22.
[52] Kurpinski K, Chu J, Hashi C, Li S. Anisotropic mechanosensing by mesenchymal stem cells. Proc Natl Acad Sci USA 2006;103:16095−100.
[53] Leucht P, Kim JB, Wazen R, Currey AJ, Nanci A, Brunski JB, et al. Effect of mechanical stimuli on skeletal regeneration around implants. Bone 2007;40:919−30.
[54] Liu J, Yu W, Liu Y, Chen S, Huang Y, Li X, et al. Mechanical stretching stimulates collagen synthesis via down-regulating SO2/AAT1 pathway. Sci Rep 2016;6:21112.
[55] Lu P, Takai K, Weaver VM, Werb Z. Extracellular matrix degradation and remodeling in development and disease. Cold Spring Harb Perspect Biol 2011;3(12):a005058.
[56] Mammoto A, Mammoto T, Ingber DE. Mechanosensitive mechanisms in transcriptional regulation. J Cell Sci 2012;125(13):3061−73.
[57] Marrese M, Lonardoni D, Boi F, van Hoorn H, Maccione A, Zordan S, et al. Investigating the effects of mechanical stimulation on retinal ganglion cell spontaneous spiking activity. Front Neurosci 2019;13:1−13.
[58] Maul TM, Chew DW, Nieponice A, Vorp DA. Mechanical stimuli differentially control stem cell behavior: morphology, proliferation, and differentiation. Biomech Model Mechanobiol 2011;10:939−53.
[59] McBeath R, Pirone DM, Nelson CM, Bhadriraju K, Chen CS. Cell shape, cytoskeletal tension, and rhoa regulate stem cell lineage commitment. Dev Cell 2004;6(4):483−95.
[60] McBride SH, Silva MJ. Adaptive and injury response of bone to mechanical loading. Bonekey Osteovision 2012;10:192.
[61] Menon S, Beningo KA. Cancer cell invasion is enhanced by applied mechanical stimulation. PLoS One 2011;6(2):e17277.
[62] Mihic A, Li J, Miyagi Y, Gagliardi M, Li SH, Zu J, et al. The effect of cyclic stretch on maturation and 3D tissue formation of human embryonic stem cell-derived cardiomyocytes. Biomaterials 2014;35(9):2798−808.
[63] Mitra SK, Hanson DA, Schlaepfer DD. Focal adhesion kinase: in command and control of cell motility. Nat Rev Mol Cell Biol 2005;6:56−68.
[64] Miyanishi K, Trinidade MC, Lindsey DP, Beaupre GS, Carter DR, Goodman SB, et al. Effects of hydrostatic pressure and transforming growth factor-beta 3 on adult mesenchymal stem cell chondrogenesis in vitro. Tissue Eng 2006;12:1419−28.
[65] Miyanishi K, Trinidade MC, Lindsey DP, Beaupre GS, Carter DR, Goodman SB, et al. Dose- and time-dependent effects of cyclic hydrostatic pressure on transforming growth factor-beta 3-induced chondrogenesis by adult mesenchymal stem cell in vitro. Tissue Eng 2006;12:2253−62.
[66] Möhl C, Kirchgeßner N, Schäfer C, Küpper K, Born S, Diez G, et al. Becoming stable and strong: the interplay between vinculin exchange dynamics and adhesion strength during adhesion site maturation. Cell Motil Cytoskeleton 2009;66:350−64.
[67] Nandi SS, Mishra PK. Harnessing fetal and adult genetic reprograming for therapy of heart disease. J Nat Sci 2015;J1(4):e71.

[68] Northcott JM, Dean IS, Mouw JK, Weaver VM. Feeling stress: the mechanics of cancer progression and aggression. Front Cell Dev Biol 2018;28:6–17.
[69] Oh S, Brammer KS, Li YSJ, Teng D, Engler AJ, Chien S, et al. Stem cell fate dictated solely by altered nanotube dimension. Proc Natl Acad Sci USA 2009;106(7):2130–5.
[70] Orr AW, Helmke BP, Blackman BR, Schwartz MA. Mechanisms of mechanotransduction. Dev Cell 2006;10:11–20.
[71] Palmisano MG, Bremner SN, Hornberger TA, Meyer GA, Domenighetti AA, Shah SB, et al. Skeletal muscle intermediate filaments form a stress-transmitting and stress-signaling network. J Cell Sci 2015;128:219–24.
[72] Papachroni K, Karatzas DN, Papavassiliou KA, Basdra EK, Papavassiliou AG. Mechanotransduction in osteoblast regulation and bone disease. Trends Mol Med 2009;15:208–16.
[73] Park JS, Chu JS, Cheng C, Chen F, Chen D, Li S. Differential effects of equiaxial and uniaxial strain on mesenchymal stem cells. Biotechnol Bioeng 2004;88:359–68.
[74] Pelaez D, Huang CY, Cheung HS. Cyclic compression maintains viability and induces chondrogenesis of human mesenchymal stem cells in fibrin gel scaffolds. Stem Cells Dev 2008;18:93–102.
[75] Prendergast PJ, Huiskes R, Soballe K. Biophysical stimuli on cells during tissue differentiation at implant interfaces. J Biomech 1997;30:539–48.
[76] Rivera-Denizard O, Diffoot-Carlo N, Navas V, Sundaram PA. Biocompatibility studies of human fetal osteoblast cells cultured on gamma titanium aluminide. J Mater Sci Mater Med 2008;19(1):153–8.
[77] Roan E, Waters CM. What do we know about mechanical strain in lung alveoli? Am J Physiol Lung Cell Mol Physiol 2011;301:625–35.
[78] Robling AG, Turner CH. Mechanical signaling for bone modeling and remodeling. Crit Rev Eukaryot Gene Expr 2009;19:319–38.
[79] Rosso G, Guck J. Mechanical changes of peripheral nerve tissue microenvironment and their structural basis during development. Bioengineering 2019;3(3):036107.
[80] Ruszczak Z. Effect of collagen matrices on dermal wound healing. Adv Drug Deliv Rev 2003;55:1595–611.
[81] Santiago-Medina P, Sundaram PA, Diffoot-Carlo N. The effects of microarc oxidation of gamma titanium aluminide surfaces on osteoblast adhesion and differentiation. J Mater Sci Mater Med 2014;25:1577–87.
[82] Schellenberg A, Joussen S, Moser K, Hampe N, Hersch N, Hemeda H, et al. Matrix elasticity, replicative senescence and DNA methylation patterns of mesenchymal stem cells. Biomaterials 2014;35(24):6351–8.
[83] Schwar US, Gardel ML. United we stand – integrating the actin cytoskeleton and cell–matrix adhesions in cellular mechanotransduction. J Cell Sci 2012;125(13):3051–60.
[84] Sears C, Kaunas R. The many ways adherent cells respond to applied stretch. J Biomech 2016;49:1347–54.
[85] SEER Training Modules. Structure of bone tissue. U. S. National Institutes of Health, National Cancer Institute, 2019. <https://training.seer.cancer.gov/>.
[86] Serrano J, Cora-Cruz J, Diffoot-Carlo N, Sundaram PA. Adaptive responses of murine osteoblasts subjected to coupled mechanical stimuli. J Mech Behav Biomed Mater 2018;77:250–7.
[87] Shieh AC, Athanasiou KA. Principles of cell mechanics for cartilage tissue engineering. Ann Biomed Eng 2003;31:1–11.
[88] Stamenkovic I. Extracellular matrix remodelling: the role of matrix metalloproteinases. J Pathol 2003;200(4):448–64.
[89] Stavenschi E, Labour MN, Hoey DA. Oscillatory fluid flow induces the osteogenic lineage commitment of mesenchymal stem cells: the effect of shear stress magnitude, frequency, and duration. J Biomech 2017;55:99–106.
[90] Subramony SD, Dargis BR, Castillo M, Azeloglu EU, Tracey MS, Su A, et al. The guidance of stem cell differentiation by substrate alignment and mechanical stimulation. Biomaterials 2013;34:1942–53.

[91] Sukharev S, Sachs F. Molecular force transduction by ion channels: diversity and unifying principles. J Cell Sci 2012;125(Pt 13):3075–83.
[92] Suki B, Sato S, Parameswaran H, Szabari MV, Takahashi A, Bartolák-Suki E. Emphysema and mechanical stress-induced lung remodeling. Physiology (Bethesda) 2013;28:404–13.
[93] Sumanasinghe RD, Bernacki SH, Loboa EG. Osteogenic differentiation of human mesenchymal stem cells in collagen matrices: effect of uniaxial tensile strainon bone morphogenetic protein (BMP-2) mRNA expression. Adv Tiss Eng 2006;12:3459–65.
[94] Talbot J, Audrey Lamora A, Stresing V, Franck Verrecchia F. Gap junction in bone remodeling and in primary bone tumors: osteosarcoma and Ewing sarcoma. Bone Cancer. 2nd ed. Elsevier; 2015. p. 83–9.
[95] Thompson WR, Rubin CT, Rubin J. Mechanical regulation of signaling pathways in bone. Gene 2012;503:179–93.
[96] Thompson WR, Scott A, Loghmani MT, Ward SR, Warden SJ. Understanding mechanobiology: physical therapists as a force in mechanotherapy and musculoskeletal regenerative rehabilitation. Phys Ther 2016;96:560–9.
[97] Topp KS, Boyd BS. Structure and biomechanics of peripheral nerves: nerve responses to physical stresses and implications for physical therapist practice. Phys Ther 2006;86:92–109.
[98] Tse JM, Cheng G, Tyrrell JA, Wilcox-Adelman SA, Boucher Y, Jain RK, et al. Mechanical compression drives cancer cells toward invasive phenotype. Proc Natl Acad Sci USA 2012;109 (3):911–16.
[99] Wagner DR, Lindsey DP, Li KW, Tummala P, Chandran SE, Smith RL, et al. Hydrostatic pressure enhances chondrogenic differentiation of human bone marrow stromal cells in osteochondrogenic medium. Ann Biomed Eng 2008;36:813–20.
[100] Wang JH, Thampatty BP. Mechanobiology of adult and stem cells. Int Rev Cell Mol Biol 2007;271:301–46.
[101] Werner M, Blanquer SB, Haimi P, Korus G, Dunlop JWC, Duda GN, et al. Surface curvature differentially regulates stem cell migration and differentiation via altered attachment morphology and nuclear deformation. Adv Sci News (Weinh) 2016;4(2):1600347.
[102] Wirtz HR, Dobbs LG. The effects of mechanical forces on lung functions. Respir Physiol 2000;119:1–17.
[103] Wolff J. Das Gesetz der Transformation der Knochen. Berlin, Hirschwald; 1892.
[104] Wolfgang H, Goldmann WH, Auernheimer V, Thievessen I, Fabry B. Vinculin, cell mechanics and tumour cell invasion. Cell Biol Int 2013;37:397–405.
[105] Wong VW, Longaker MT, Gurtner GC. Soft tissue mechanotransduction in wound healing and fibrosis. Semin Cell Dev Biol 2012;23:981–6.
[106] Yang PF, Brüggemann GP, Rittweger J. What do we currently know from in vivo bone strain measurements in humans? J Musculoskel Neuron Interact 2011;11:8–20.
[107] Yi E, Sato S, Takahashi A, Parameswaran H, Blute TA, Bartolák-Suki E, et al. Mechanical forces accelerate collagen digestion by bacterial collagenase in lung tissue strips. Front Physiol 2016;7:287.
[108] Zhang H, Labouesse M. Signalling through mechanical inputs: a coordinated process. J Cell Sci 2012;125(Pt 13):3039–49.
[109] Zhang W, Kong CW, Tong MH, Chooi WH, Huang N, Li RA, et al. Maturation of human embryonic stem cell-derived cardiomyocytes (hESC-CMs) in 3D collagen matrix: effects of niche cell supplementation and mechanical stimulation. Acta Biomater 2017;49:204–17.
[110] Zhou S, Salisbury J, Preedy VR, Emery PW. Increased collagen synthesis rate during wound healing in muscle. PLoS One 2013;8:e58324.
[111] Zhu C. Mechanochemitry: a molecular biomechanics view of mechanosensing. Ann Biomed Eng 2014;42:388–404.

Index

Note: Page numbers followed by "*f*" and "*t*" refer to figures and tables, respectively.

N

O

P

R

S

Kolibri 360 GmbH
Ferdinand-Jühlke-Straße 7
99095 Erfurt, Deutschland
produktsicherheit@kolibri360.de